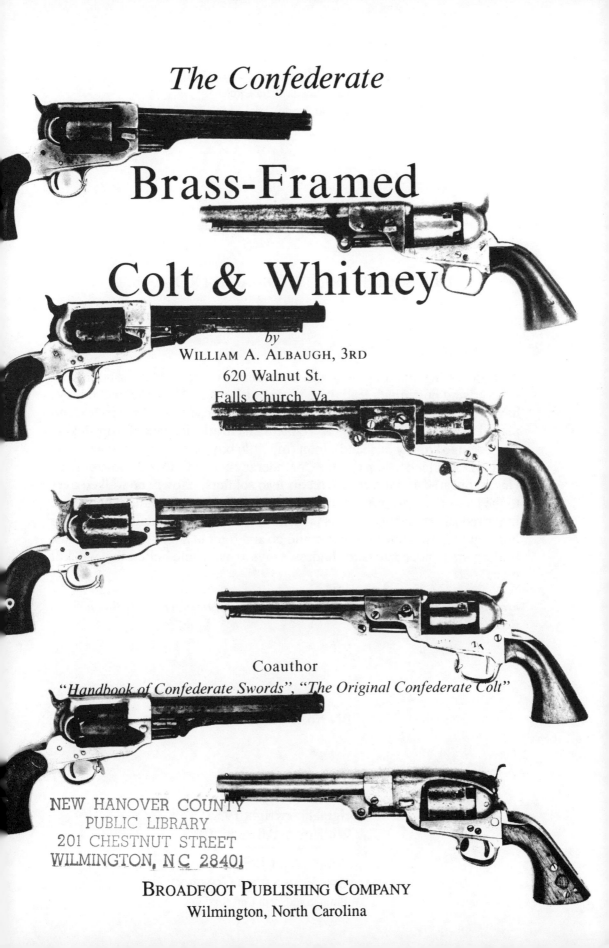

# The Confederate Brass-Framed Colt & Whitney

*by*
WILLIAM A. ALBAUGH, 3RD
620 Walnut St.
Falls Church, Va.

Coauthor
"*Handbook of Confederate Swords*", "*The Original Confederate Colt*"

BROADFOOT PUBLISHING COMPANY
Wilmington, North Carolina

*"Fine Books Since 1970."*
BROADFOOT PUBLISHING COMPANY
1907 Buena Vista Circle
Wilmington, North Carolina 28405

THIS BOOK IS PRINTED ON ACID-FREE PAPER

"We are saving lead to give to the Confederate Government for bullets. It is surprising how much lead there is about a house. All of the lead pulleys have been taken from the windows as well as the lead around the panes, and placed in a large clothes basket. To them I have added a box of lead sinkers that belonged to my husband. John (my little boy) asked what the lead was for and I explained to him that the Confederacy needed it to make into bullets. He disappeared and returned with his lead soldiers. Slowly he walked over to the basket and dropped them in. I rescued one, the Captain, who wore a brave red jacket and was John's favorite. I told him the Captain could be the home guard. The other soldiers could go and fight in a real battle. We must be desperate, indeed, to take children's toys to make into bullets that will kill men."

*From the diary of Caroline Jamison Jenkins, wife of Gen. Micah Jenkins, C. S. A. of South Carolina (August 15, 1864)*

ISBN No. 1-56837-265-5

Original copyright 1955
by William A. Albaugh, III

Copyright 1993
BROADFOOT PUBLISHING COMPANY
Wilmington, North Carolina

*To my elder son
Christopher Skipwith Albaugh
whose initials spell, "C. S. A."*

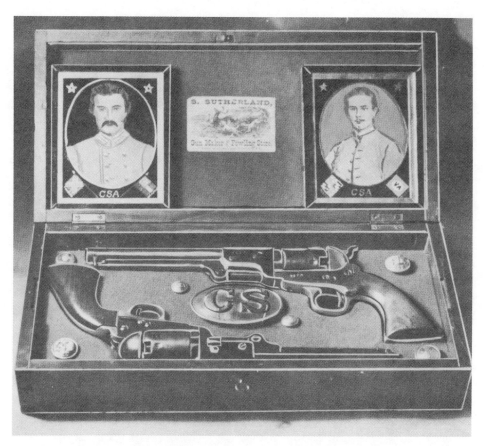

A fine pair of Confederate Revolvers. Cased after the war. Griswold & Gunnison #2419, and Leech & Rigdon #693.

Pictures of Lt. John William Albaugh, killed at Bunker Hill, Va. 1864, and Sgt. Ira Albaugh, killed at Kelly's Ford, Va. 1863. Both of the First Virginia Cavalry, C. S. A. *(from the collection of the author)*

# TABLE OF CONTENTS

INTRODUCTION
PART 1—The Griswold & Gunnison Revolvers
PART 2—The Spiller & Burr Revolvers
APPENDIX

# LIST OF ILLUSTRATIONS

| | |
|---|---:|
| Cased Griswold & Gunnison and Leech & Rigdon | *Frontispiece* |
| Revolver parts found at Griswoldville | 8 |
| Griswold & Gunnison #608 with round barrel housing | 10 |
| A Reconstructed Rebel | 12 |
| Sword made at Macon, Ga. | 14 |
| Right and left side of Griswold & Gunnison #2198 | 16 |
| Four Spiller & Burrs, and 4 Griswold & Gunnisons | 24 |
| Right and left side of "Sample" Spiller & Burr | 41 |
| Map of Wartime Atlanta, Ga. | 62 |
| Markings on a Spiller & Burr barrel | 82 |
| Cased Spiller & Burr #983 | 88 |

# INTRODUCTION

The blades of Madrid, Spain have a well earned reputation for flexibility. None the less flexible was the policy of the Confederate Ordnance Bureau, which was able to exist only through its ability of borrowing from Peter to pay Paul. Substitutions were relied upon for practically all essentials, and in many cases after the original supply of substitutions had been used, the hard pressed, but ever ingenious Bureau was forced to rely upon a substitute for the original substitution.

Thus it was that iron originally was used in lieu of steel, the iron to be replaced by brass, and in turn, pewter and lead took the place normally occupied by brass.

Brass church bells, copper turpentine stills were melted down to be turned into hilts for swords, and revolver frames. Lead roofing and window weights went to make bullets and coat buttons, while kitchen knives and similar articles were used for their steel content.

Little wonder it is that Confederate Ordnance has so long appeared to the historian as a jig-saw puzzle—jumbled, and with many pieces missing. Through the research though of this person (such as R. D. Steuart) and by that person (such as L. D. Satterlee), the overall pattern is gradually becoming apparent, and only the details are lacking.

Many of these previously missing details are to be found in our National Archives in Chapter IV of the "Captured Rebel Ordnance Records." Reference in this work to "volume" refers to the various volumes in this series which exist only in their original record form.

A former monograph "The Original Confederate Colt," told the story of the Leech & Rigdon, Rigdon-Ansley revolvers. This present work will deal with those handguns whose frames were made of brass, and are logically known to the arms collecting brotherhood as being of the "brass-framed" variety.

Despite its obvious answer, many theories have been advanced as to why the Confederates made such extensive use of brass in practically all their weapons. There can be only one answer—brass was easier to work than iron, and thus needed less complicated machinery, and less skillful workmen in its handling. We shall also see how iron was substituted for steel, although such usage is not quite so apparent visually.

In my files is a personal letter from Lt. Col. J. W. Mallet, former C. S. Superintendent of Ordnance Laboratories. Although Col. Mallet only touches on the subject at hand, his letter is never the less of general interest, and so is quoted.

"University of Virginia
Charlottesville, Va.
June 8, 1909

My Dear Sir:

Your letter of the 6th inst. reached me this morning, and I wish I could help you to an answer to your questions, but the large part of my papers and memoranda, letters, of Civil War times, was lost by fire in 1884, and I find increasing difficulty in drawing upon my memory for facts in detail.

I believe that the names you mention as stamped upon small-arms are those of private gunsmiths who did the work for individuals, and also in most cases for the Confederate Government through Ordnance Officers at different points. In one case I known this is so—S. Sutherland was the leading, possibly the only gunsmith in Richmond at the outbreak of the war, and he did a large amount of work both privately and for the C. S. Govt. In many instances, weapons were repaired using the parts of others—this may be the case with your gun marked 'Tyler, Texas.' It is I believe an established historical fact that John Brown brought into Virginia some rather crude pikes with which he intended to arm the negroes for an expected servile insurrection, but I do not know where they were made. Some at least of the foot artillery swords were made in Richmond in the very early part of the war. The extensive use of brass instead of steel, to which you refer, was due to the lack of skilled smiths, and the easier manipulation of brass by casting. I remember my horse running away with me at Yorktown in consequence of a cast brass bit breaking in his mouth when he made a sudden plunge at the bursting of a heavy gun in one of the works near by.

Regretting that I cannot give you fuller information, I am

Very truly yours,
J. W. Mallet."

The boom of Sumter's guns in 1861 found the seceding Southern States without a single small arms factory of any size within its borders. We are excluding the Virginia Armory in Richmond, which was being retooled, and the Harper's Ferry Armory & Arsenal which was Federal property. However, all Yankee ingenuity and enterprise was not north of the Potomac, and numerous plants for the manufacture of swords, pikes, bayonets, muskets, rifles and pistols were started in the South by both individuals and the C. S. Govt.

The Confederate Congress early anticipated the need for weapons and other munitions of war, and various acts were passed, very favorable towards those who would supply the implements with which to fight. These inducements included: cash to the extent the manufacturer needed no money of his own to proceed, only the promise and guarantee that this outlay of monies would be returned. To make this even more attractive, the loan was made without interest for a certain period of time. Still in addition, the Govt. promised that the manufacturer would make money from his enterprise, and that in the event of anything so unfortunate as his plant falling into the hands of the Yankees, would be repaid lock, stock and barrel, and the initial loan would be cancelled. Still further, those engaged in arms-making were exempt from military service.

To this generation with memories of World War #II still fresh, and reminiscent of "cost-plus," "escalator clauses," "exemption for riveters" etc., the above is not unusual, but in the middle 1850's, such a step was revolutionary.

The inducements were so attractive that a number of persons immediately applied for Govt. grants, which at first were given rather freely. Contracts for 5,000, 10,000, and even 25,000 arms were signed without the blinking of an eye by persons totally unqualified to envision the vicissitudes of manufacturing under conditions of stress.

Many of these ventures were futile almost to the point of pathos, such as Holly Springs, which was an utter fiasco. Others made a model of their proposed weapon (either in fact or on paper), and then quietly passed out of the picture.

Some, however, like the Enfield gun plant of Cook & Bros., the various Haiman industries, were planned and carried out with such pains and the expenditure of so much brains and capital that it seems a pity they were so short-lived. They were closed by one factor only—the enemy.

Also on the plus side of the ledger are such names as: Leech, Rigdon, Griswold, Gunnison, Spiller and Burr, and we are more concerned with these names rather than the "pilot models" which still turn up and which indicate at best only a very small output. A later work will attempt to cover some of these so-called "pilot models" but first let us get to those men who were the real meat and potatoes of Confederate revolver making.

It has already been mentioned that in the early stages of the war, contracts were accepted for as much as 15,000 revolvers, but in fact, through the entire four years the South struggled for its existence this amount was not produced by the combined efforts of all its pistol makers. Note that I use here "pistol-makers" and include single shots as well as revolvers. It is extremely doubtful that as many as 10,000 pistols and/or revolvers were ever made within the confines of the Confederacy.

To break this figure down; Griswold & Gunnison's total output was slightly over 3,600; Leech & Rigdon—Rigdon Ansley, less than 2,500, while Spiller & Burr's total approached 1,500. The above were the largest producers in the South, and totalled less than 7,500 revolvers. Other manufacturers were unable to produce much more than 1,000.

A book such as this, is obviously not the product of one man. Someone has said that a historian is only one who puts in words what other people have already said or done, and in this case this is certainly true. Originally, Richard D. Steuart and I planned to do this series of monographs together. His death prevented this. While he had no actual part in this present work, I am indebted mainly to him for the groundwork and notes he left behind, and which I have not hesitated to use. I am also indebted to the countless collectors with whom I have corresponded, each one of whom has added and contributed to the whole. To them, I give a sincere thanks, and regret that each can not be mentioned by name. In particular, however, I would like to acknowledge the assistance from my following friends:

John Amber, Chicago, Ill.; Charles Benton, Macon, Ga.; Richard Brady, Baltimore, Md.; Miss Eleanor Brockenbrough, Richmond, Va.; Harry H. Brooks, Dallas, Texas; Oscar DePrato, Wash., D. C.; Mr. & Mrs. Joe Etheridge, Macon, Ga.; W. W. Eubanks, Concord, N. C.; Victor Friedrichs, Austin, Texas; Foster Gleason, Washington, D. C.; Walter Goldstein, New Orleans, La.; William Harden, Augusta, Ga.; C. C. Holloway, Longview, Texas; S. L. Hutcheson, Greenwich, Conn.; Col. Catesby Jones, Richmond, Va.; Leon C. Jackson, Dallas, Texas; Harry Knode, Dallas, Texas; William M. Locke, Cincinnati, Ohio; Walter Massie, Macon, Ga; Robert Moates, Richmond, Va.; C. Meade Patterson, Wash., D. C.; Andrew Palmer, Dearborn, Mich.; Harold Peterson, Wash., D. C.; Claude E. Petrone, Washington, D. C.; W/O J. L. Rawls, Philadelphia, Pa.; Jerry J. Reen, Alexandria, Va.; Frank Russell, Brainerd, Minn.; Herman Schindler, Charleston, S. C.; James E. Serven, Santa Ana, Calif.; Edward N. Simmons, Kansas City, Mo.; Fred Slaton, Jr., Madisonville, Ky.; Sam Smith, Markesan, Wis.; W. Thomas Smith, Richmond, Va.; James R. Somers, Dallas, Texas; Herman Strumpf, Cincinnati, Ohio; Miss Bertha Thompson, Richmond, Va.; J. S. White, Highland Park, Ill.; the late Ben Ames Williams; Hermann W. Williams, Jr., Wash., D. C.; M. Clifford Young, Boston, Mass.; the American Rifleman, Washington, D. C.; the Battle Abbey, Richmond, Va.; the Texas Gun Collector, Dallas, Texas; The Texas State Library, Austin, Texas and the Virginia Historical Society, Richmond, Va.

# PART I
# The Brass-Framed Confederate "Colt," A Story of the Revolvers Made at Griswoldville, Ga.

# GRISWOLDVILLE, GA.

GRISWOLDVILLE, FORMER STATION ON THE Central of Georgia Railroad, is situated some 10 miles southeast of Macon, Ga. There is no need to look for it on the map. It existed as a post office only during the time the mail was delivered by the Confederate States Government.

Maps prior to 1860 occasionally show a dot marked "Griswold" or "Griswold Station." During the war period it proudly bore the name "Griswoldville." After 1865, its name reverted back to "Griswold." It could never have been very large. The census of 1880 shows Jones County, Ga. in which it is located, to have had only 3,753 white persons living in the entire county!

## GRISWOLD IN 1953

Today the place is hard to find. It is located in an extremely isolated spot, five miles from the nearest paved road and consists only of 4 or 5 widely scattered houses. There is no store, or filling station and it is literally an intersection of two narrow dirt roads along a railroad whose trains no longer stop, and for which there is no station. One deserted house stands at the southeast corner of this intersection. This house, about 50 years old, is built on the foundations of the old Griswold home which burned some years after the war. A mile or so to the west of what was once the town, lie the brick foundations of a large building. This in 1861-65 had been a soap manufactory.

The revolver factory stood near the northwest corner of the intersection, on the north side of the railroad. The site is now an empty field which has been plowed and replowed many times. No trace of the factory remains although a life-long resident pointed out where the old foundry chimney had stood, it being all that was left of the plant after it was destroyed by the Yankees in 1864.

## PIECES OF REVOLVERS ARE STILL BEING FOUND

Spurred by the account of a nearby farmer having plowed up various revolver parts, several hours were spent closely examining the area. The search was rewarded by kicking up the female die from which were struck the rough hammer blanks. Being of fine tool steel, it is in remarkably good condition for having laid in the ground for 90 years. This prize was well worth the hours spent.

Iron revolver parts are still being found in the vicinity of Griswoldville, and thanks to Mrs. Joe Etheridge and Charles Benton, I have in my possession a number of barrels, cylinders, hammers, triggers, cylinder pins, etc. (See Photo.) Most appear to be discards, being imperfect in one way or another. Others, however, seem perfect—only unfinished. These last indicate having been exposed to intense heat or fire, and might possibly be parts which were in the process of being finished when the factory was burned in November 1864!

Brass parts are rarely found. It is assumed that if a barrel or cylinder was improperly bored the piece was tossed onto the scrap pile (later to be found), but an imperfect brass part was probably thrown back into the brasspot to be melted and recast.

Speaking of the brasspot, while at Griswoldville, I was shown where "a large cast iron pot full of brass" had stood for many years after the war, finally to be rolled away and cast into an unused well "to get it out of the way." Undoubtedly it was this "pot" that contained the brass used for the frames, trigger guards etc.

An enterprising young man of this section has mounted on a board the various revolver parts that he personally has uncovered at Griswoldville. At this point he has a complete revolver except for the backstrap and grips. I hasten to add, however, that all such parts are either unfinished, or obvious discards, and their original poor condition has not been improved by laying in the ground so long. The cylinder is in marked contrast to the rest of the parts, being in reasonably good condition, and on which the serial #1144 is plainly evident.

I believe it will come as a distinct shock to collectors to learn that the frame of this "made-up-piece-by-piece" revolver is not brass at all. It is iron! It is badly warped and never could have been finished, and is obviously a discard. Nevertheless, the fact remains that it is of iron, thus proving beyond doubt that an attempt was made to manufacture these revolvers with iron frames. Whether this attempt was successful or not I am not prepared to say, but should an iron-framed Griswold Colt appear one of these days, it might be well to examine it closely before declaring it a fake.

Despite today's air of desolation, in the 1860's, Griswoldville was a spot of much activity, for here were made during three hectic war years the famed "Confederate Brass-framed Colt," prized by all gun collectors as a very typical Confederate weapon from the tool-marked barrel to the shiny red or yellow brass frame. Some 3,600 of these revolvers were made, and at the highest point of production 5 per day were counted as completed, or 100 per month.

## DESCRIPTION OF THE REVOLVER

Except for an unimportant deviation, all known Griswold "Colts" conform to type. All are .36 calibre, 6 shot, brass-framed imitations of the 3rd model Colt dragoon. All had the round barrel, and most had the dragoon-type part octagon barrel frame, although the barrel frame housing was round in the early model. (See Photo.) This last is the only distinguishing difference between any of them for otherwise they are as alike as any handmade items can be. Reference is made now to appearance and not markings which were of course different on each gun.

*Barrel*—7½ inches round, tool marks plainly evident.
*Rifling*—6 lands and grooves right with a gain twist (compare this with a Colt or Leech & Rigdon).
*Barrel frame*—round or part octagonal. 1-60/64 inches long.
*Foresight*—brass pin.
*Barrel wedge*—made without spring. (Most contain secondary serial number.)
*Loading assembly*—closely resembles that of a Colt navy, with navy type catch.
*Cylinder*— made of *twisted iron* with the twist visually evident. 1¾ inches long, diameter 1-9/16.
*Safety device*—pins between cylinder cones, similar to Colt.
*Frame*—red or yellow brass, no capping channel, tool marks evident.
*Trigger guard*—brass, red or yellow, with tool marks evident.
*Backstrap*—brass, red or yellow, with tool marks evident.
*Hammer*—provided with roller.
*Grips*—walnut, evidently not properly aged, for in most cases those surviving guns are found with grips too small, believed to be due to the shrinkage of the wood throughout the passing 90 years. The full serial number of the gun is usually found pencilled on the grip under the backstrap but in some cases, the secondary serial is stamped on the butt.

*Overall length*—about 13 inches. Handle of the weapon at greater angle than Colt, giving the impression that the gun has been used as a club.

The measurements are so nearly the same on known guns that one wonders why collectors believed for years that the brass-framed "Colts" were the products of many factories or plants throughout the South, rather than having all come from one place? This erroneous theory was brought to an abrupt end with R. D. Steuart's article "The Confederate Colt" which appeared in "Army Ordnance" September 1934, and collectors now know they were all made at Griswoldville, Ga.

## GRISWOLD & GRIER

Ever since it has been established that all these revolvers came from one point alone, they have been known as "Griswold & Griers". and are rarely referred to in any other fashion. At this late date I hesitate to change this now established name, but will say that there is no official correspondence which refers to them thus. Officially, that is to Confederate Ordnance officers, they are referred to only as "Griswold & Gunnisons", and that is what they should be called.

The name "Griswold & Grier" was probably given to this revolver by E. Berkley Bowie, the first serious Confederate collector, who through his research laid the ground work for much present day general knowledge.

A letter now in my possession, dated March 13, 1923, from Mrs. Ellen Griswold Hardeman, of Macon, Ga., niece of Giles G. Griswold, to E. Berkley Bowie, of Baltimore, Md., says in part:

"......My uncle, Giles G. Griswold (in person) went to Montgomery, Ala. to negotiate the leasing of my grandfather's large plant (which had prior to that time manufactured the Griswold Cotton Gin, and sold by agents throughout the Southern States). President Davis and his cabinet in Montgomery leased the plant in 1861, or thereabout. My grandfather, an old man was retired from business, and as my uncle Giles, died in Columbus, Ga., while returning from Montgomery, my uncle-in-law E. C. Grier took charge of the entire business. As all the shops, foundry, etc., were destroyed by Sherman's army in their ravaging march from 'mountain to seashore' and a battle at Griswoldville ensuing, we opine that those old firearms (large oldfashioned pistols) were destroyed or captured by the Yanks.

I was a 'miss of sixteen' when Sherman's army passed thru and 'laid in ashes' our dear old Griswoldville, leaving only my grandfathers' and the Grier home standing."

The above letter is quite clear, and we can not wonder that Bowie placed the "tag" of "Griswold and Grier" upon the product of Griswoldville.

## SAMUEL GRISWOLD

Records, however do not quite agree with Mrs. Hardeman. "Georgia Landmarks, Memorials and Legends", page #712 says the following:

"Samuel Griswold, and Daniel Pratt, two ingenious and wide awake pioneers came to Clinton, Ga., from the State of Conn.......and in a modest way began to build cotton gins.......Over 900 of these gins were sold annually..... Mr. Pratt removed to Alabama, but Griswold established the town of Griswoldville. The iron works there were so destroyed by the Federals they were never rebuilt."

Samuel Griswold, founder of Griswoldville, is evidently the "grandfather" referred to by Mrs. Hardeman. So far so good, and the Georgia Landmarks goes to bear out her letter.

Mrs. Hardeman says however; "My grandfather, an old man, was retired from business." Samuel Griswold may have been old, but as for being retired from business, that is another matter. Possibly he had retired from active business, but came out of retirement in short order when his Conn. Yankee instincts saw prospects of War contracts for the Griswold Cotton Gin Co.

Another account of Samuel Griswold, and which dovetails nicely with those already cited, was furnished through the courtesy of Mrs. Tommy Wells, principal of Griswold School, who made available a letter from Mrs. Claude B. Wilson, formerly Martha Van Buren, another granddaughter of S. Griswold. Writes Mrs. Wilson·

"Samuel Griswold, for whom Griswoldville was named was born in Windsor, Conn., Dec. 29, 1790. He came to Clinton, Jones County, Ga. with his wife, Lois Forbus Griswold and son Roger in about 1814.

Several years later he erected a large machine and cotton gin ship where he began to manufacture the famous Griswold Gin for ginning cotton. This manufactory grew to such an extent that about 1835 Mr. Griswold purchased a large area of some 5,000 acres of land in the town section of Jones County, about 10 miles south of Macon.

The Central of Georgia Rwy. from Macon to Savannah had just been completed and ran through this plot.

Here, Mr. Griswold erected a gin factory, saw and grist mills—a large brick foundry and planning mill were also built and modern machinery was supplied the enterprise.

He also built a fine three storied residence with barns and stables.

Across the railroad track (on the factory side) he erected a post office and 50 or more cottages of 4 and 6 rooms for his white employees and his negro house servants.

He also built a handsome church, and a large commissary which contained an ample supply of food products, most of which were raised on his farms.

Attractive homes were also built for his 3 sons, his 5 daughters and head factory foremen.

He had a number of salesmen who traveled through Alabama, Louisiana, Mississippi, Arkansas, Kansas and Texas, taking orders for the Griswold Gin. The gins ordered were delivered by wagon as railroad facilities were not available in most sections.

Everything used by them in the manufacture of the gins was made right at Griswoldville.

Soon shipping facilities became better, sales increased and thousands of Griswold Gins were in use in every cotton growing State.

Mr. Griswold owned about 100 negro slaves, and was a most generous and liberal master. After the war some of these slaves preferred to remain with their old master rather than to accept their liberty.

He kept his enormous fortune in two big iron safes—one in the Post Office, the other in his private sitting room at his home.

In 1862 the large cotton gin plant was leased to the Confederate Government for manufacturing army pistols.

When Sherman made his memorable 'march through Georgia,' a battle was fought here and all the vast properties were destroyed by fire. Only the Griswold residence and 2 others were untouched. Why these were left, one can not explain.

Prior to 1862, Mr. Griswold was rated as one of the richest men in Georgia. He died in Jan. 1869."

Taking the 3 cited accounts, plus what we know from other sources, we can guess that with the war, the retired Sam Griswold came out of retirement and sent "Uncle Giles" packing to Montgomery while he resumed charge of the business, and scratched around for what local contracts he could pick up. The mention that Giles had been sent to Montgomery indicates clearly this was in the early days of the war, as that city was the Confederate Capitol for only a short time before it was removed to Richmond, Va.

## GEORGIA PIKES

The war found the individual States as well as the Confederate Government sorely in need of weapons, and on Feb. 20, 1862, Gov. Joe Brown of Georgia appealed to the patriotism of mechanics and manufacturers to furnish the State with "pikes on a 6 foot staff." Backing up the appeal to patriotism was the added inducement of $5.00 for each pike furnished.

Evidently farsighted enough to know that cotton gins would hardly be in demand at this particular time, Sam Griswold converted his plant to the manufactory of pikes.

This conversion was not too difficult a job when we realize that in those days most machinery was termed "all purpose," and was not of the specialized variety of the 1950s.

The "Confederate Records of Georgia," Vol. #11, page 352, shows the number of pikes received by the State of Georgia, listing each manufacturer, number and date received.

April 3, 1862, little over 6 weeks after Gov. Brown's appeal, it is noted that Samuel Griswold sent in 16 pikes. Evidently he was just getting started. April 15th shows 90 pikes, and one week later, April 22nd—100! Sam was getting into the swing of things. Further supply dates list May 17th—300, May 27th—97, June 2nd—201, or a grand total of 804 pikes. For two months work at $5.00 per, this was not a bad piece of business.

Arms collectors may be interested to know that several of Sam Griswold's pikes have survived the years. All seen by the writer have been of the "cloverleaf" variety, with 6 foot tapering pole, round knob on butt, 3 pointed blades of which the center one is 10 inches long and pear shaped. The side blades are 3½ inches each from center to point. The blade is stamped "S. Griswold." Such a pike is in the Battle Abbey Collection, Richmond, Va.

The State records of pikes received, continued all through the year 1862, but Sam Griswold's name does not appear in this connection after June 2nd.

From what we know of Sam, he was too good a business man to let such a profitable undertaking cease from lack of interest, and it must be that this date marks the time when his activities were focused entirely on revolver making. The records bear us out on this.

## E. C. GRIER

Also in the pike making business we note a Grier & Masterson, who on

March 28th turned out 80 pikes and 33 on April 10th. There is no further mention of Grier & Masterson after this date, and we wonder if this could have been the same Grier who evidently had some connection with the revolver works?

Mrs. Hardeman's letter indicated the Grier in question was E. C. Grier, and that he was a son-in-law of old Sam Griswold.

Records are very much silent on Grier's part if any in the revolver enterprise. He is not mentioned.

Local tradition has it that Grier came to Georgia from Love or Coldwater, Miss. Some say that he was a gunsmith.

A Mr. Charlie Kitchen, born at Griswoldville in 1872, recalls Mr. Grier, having helped his father retrieve doves for that gentleman during many hunts. Mr. Kitchen further recalls that Grier lived in a 40 room house with his 3 sons, Clark, Tom and Ross. He remembers that Grier was always addressed as "Colonel" and that he always wore a beaver hat.

How Col. Grier received his title is not known. Confederate Army records do not indicate that he was a commissioned officer in that Army. Possibly he was a colonel in some State Militia, or possibly the title was merely an honorary one.

Other old residents of Griswoldville think of Col. Grier as a rather strange sort of chap—a dreamer and star gazer who used to lie out on the rocks and watch the moon and the stars, and who sometime after the war moved to Savannah, where he founded the Grier Almanac.

## MACON, GA., SPRING 1862

It now becomes necessary to introduce another character into the story. This is Capt. Richard E. Cuyler, who in 1861-62 was in charge of C. S. Ordnance at Savannah, Ga.

Spring of 1862 brought threat to Savannah from the sea, and fearful lest the valuable ordnance stores there fall into enemy hands, Col. Gorgas, Chief of Confederate Ordnance, ordered their removal to Macon, Ga., 183 miles inland (Vol. #36, 4-25-62).

Later, Macon was to become one of the largest Confederate Ordnance centers but early 1862 found no Government Establishments there.

Capt. Cuyler arrived in Macon around the 1st of May 1862 and purchased the old Findlay Iron Works, lock, stock and barrel, even to the mechanics and timekeeper, and went about setting up the C. S. Arsenal, as he had been instructed to do. (Vol. 36, April 28, 1862) He also took over the establishment of D. C. Hodgkins & Sons, gunsmiths. (Vol. 36, May 27, 1862) Walter Hodgkins, one of the sons, was to become one of his most valued employees, and it was he who later was responsible for the inspection of the Griswoldville revolvers.

## C. S. ARSENAL, MACON, GA.

I am fortunate in having in my files a letter from one of the former operatives of the Macon Arsenal, written in May 1922 by Bridges Smith, who was then Judge of the Bibb County Juvenile Court, Macon, Ga. Says the Judge:

"In 1861 when the new Confederate Government saw the necessity of beginning at once the manufacture of army equipment, it depended largely upon patriotism (which wasn't lacking in a town like Macon) for co-operation. D. C. Hodgkins & Sons, turned over their gunsmith shop, and then began the making of pistols and conversion of old weapons into, at that time, more modern guns.

At first these were made of the Colt pattern by Hodgkins, but later were made by the Govt. of the same general Colt model, but by mechanics brought here from Harper's Ferry, and I understand the model was altered in some way. Fom 1861 to 1865, I was detailed to make ammunition for shotguns, muskets, and rifles, but none for small arms, and therefore had little opportunity of seeing the manufacture of pistols.

W. J. McElroy & Co., tinsmiths, turned over their large factory for the purpose of making canteens, and later by reason of having some skilled men in their employ, began to make swords. In the course of time, this shop was turning out some of the finest weapons of this class, beautifully ornamented by the then process of dipping the blade in melted wax and chasing the designs with a steel pointed instrument, and then pouring acid all over and letting it 'eat' into the blade. This may have been very crude compared to modern methods, but we used to regard the product as almost perfection. (See photo.)

At any rate, I am satisfied that these were the first swords made and worn by the officers of the Confederate Army. This factory continued until the war ended and the swords were made into plowshares, as the saying is. The government had no sword-making establishment of its own here.

The first cartridges made here, and about the first made in the South, were of the 'buck & ball' variety for shotguns, composed of one large (about the size of a boy's marble) leaden ball and 3 buckshot, a thimble-full of powder, all in a container of brown wrapping paper and tied with small twine, the powder end being folded in a way that would hold its position. This is the cartridge that brought the command 'chaw cartridge,' following after the command 'load.' The soldier bit off the folded end and poured the powder in the barrel of the gun, and then put the buck and ball with their paper wrapper in, the paper serving as the wadding, and then rammed it down with the ramrod. He then put on his percussion cap and was ready for the command 'fire.'

You can see how much time was wasted in the earlier stages of the war, when you think of the metal cartridges that came later, especially with the opposing army. Another cartridge was that of the slug for rifles. At first these were moulded, and I think of them when I read about the dum-dum bullet, both being of soft lead. Later, through the ingenuity of a Michigan man, a civil engineer and by the way the major of my battalion, these balls or slugs were made as hard as iron. He had lead wire, about the size of your little finger made in coils, and invented a machine to cut it in proper lengths and made to fall into a recess the shape and size of the slug, where by powerful pressure it was compressed into hardness.

Brown paper was also used for making these rifle cartridges until a vessel loaded with a quality of paper something of the appearance and texture of what we call 'bond-paper' used for stationary, ran the blockade at Wilmington, and then we made decent looking cartridges. After the rifle cartridge was made, the slug end was dipped in wax. All cartridges were put in packages of ten, with a small package containing 13 percussion caps.

In 1865, the Confederate Govt. had just completed large and handsome buildings, one for a laboratory in which to make ammunition, and the other for an armory in which to make small arms. Before they could be turned over by the contractors, Gen. Wilson, U.S.A., had the audacity to walk in with his cavalry and demanded the surrender of the city, and the people at that time just didn't have the heart to refuse him. After that, I lost interest in the manufacture of cartridges and weapons."

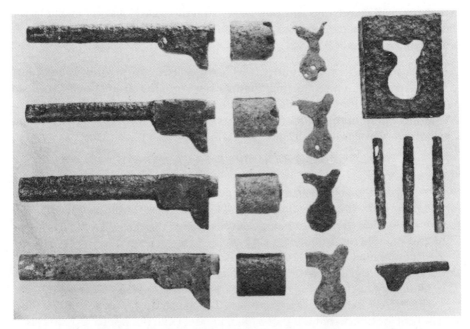

Revolver parts dug up at Griswoldville, Ga. *(from the collection of the author)* Close examination of the picture discloses various stages of manufacture. Note "female" hammer die in upper right hand corner.

On Sept. 10, 1861, Gorgas issued the following circular: "Officers commanding Arsenals, and Ordnance Depots, are expected in the present state of supplies to exert themselves individually and to the utmost to procure whatever is necessary to furnish without waiting instructions from this office.

They will therefore endeavor to purchase and contract for arms, swords, pistols, artillery and cavalry equipment, and materials for powder in anticipation of the calls that may be made on them, reporting to this office their operations." (Vol. 140)

## GRISWOLD & GUNNISON

After locating in Macon, it is probable that one of the 1st things Cuyler did was to seek contracts for arms, accoutrements, etc. Less than a month after his arrival he wrote Gorgas concerning the proposed revolver manufactory at Griswoldville. We do not have this letter, but we do have Gorgas's reply which is dated May 19, 1862:

................. "Capt., You are hereby authorized to receive from Messrs Griswold & Gunnison, all the revolvers they can make in eight months, and pay $40.00 for each-complete; they must stand the usual tests, and be of good workmanship." (Vol. #36)

Cuyler passed this information on to Griswold, to which he received this reply:

....... "Dear Sir, In answer to your communication of yesterday would say that we shall not be able to deliver you more than a specimen of our revolvers within the next two months. After then, we expect to finish and deliver between 50-60 per week regularly. As required by our contract we shall endeavour to make an article that will stand your thorough tests, and of as good workmanship as the hands we have and can procure can give them. Very respectfully, Griswold & Gunnison." (Vol. #36)

After reading so many contracts which promised any amount of anything, to be delivered at anytime starting tomorrow, we find the letter of Sam Griswold particularly refreshing. Sam makes no promises except that he will do his best.

A most illuminating document comes to us in the form of a letter to Capt. Cuyler from Walter Hodgkins dated July 16, 1862.

### THE REVOLVER FACTORY

"Dear Sir:-

"According to your instructions I repaired to Griswoldville yesterday noon for the purpose of showing defects in Colts Pattern Pistol exhibited to you, and to give such other information as would be necessary to make the pistols efficient and suitable for ordnance purposes.

After making the necessary comparisons of the components I proceeded to examine their mode of construction. I found 22 machines worked by 24 hands, 22 of whom are negro slaves. They have about 100 pistols in progress.

The barrels are forged from ordinary 1 inch square bar iron. The cylinders are cut from ordinary round bar of sufficient size. I can not approve of this process, but the proprietors feel confident that they will stand the required test. At their request I have brought up barrels and cylinders which I will subject to severe proof with your permission.

I demonstrated to the foreman the process of case hardening, tempering springs, and tinning steel without the use of acids and described the manner of blueing. I also enjoined on them the importance of high polish on the inside of the barrel to prevent the ball slipping and reducing the liability to foul.

I think they will require the services of more practical mechanics particularly for the assembling and also think it will be necessary to subject each pistol to proof, and close examination." (Vol. #36)

From the above we know that on July 16, 1862, Sam Griswold had about 100 pistols underway, that he had a factory of 22 machines operated by 24 hands of which 22 were negro slaves! This then was the original plant setup under which they hoped to turn out 50-60 revolvers a week, and most amazingly, they did!

### PROVING THE BARRELS AND CYLINDERS

July 18th, Hodgkins set about the "proving" of the barrels and cylinder, and on this date wrote to Cuyler,

.............................."Sir, By your direction I proceeded this morning to prove the Colts pattern cylinders and barrel. I had prepared false breech and clamps (weight 4 pounds), for the barrel which rested upon the ground. 1st Charge, 55 grains rifle powder, 2 round balls, and 2 wads. Results favorable, but discovered opening between breech and barrel, causing windage on top of resistance.

2nd Charge, 82 grains rifle powder, 2 conical balls and 2 wads. Result—bursted. The cylinder rested upon an iron ring (to avoid the ratchet) perpendicular. Each chamber stood the proof of 27 grains and one ball. When we found it impossible to fire the whole together, and to cause some additional resistance,

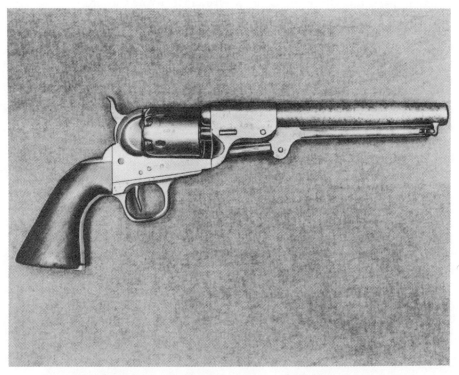

Early model Griswold & Gunnison #608 with round barrel housing *(from the collection of Edward N. Simmons)*

we placed 2 pounds of weight upon the ball and top of chamber. 1st chamber stood, 2nd—bursted.
I present herewith both cylinders and barrel that the nature of the iron may be properly inspected. I will also state the English proof charge is 120 grains Provisional (before the barrel is turned), 75 grains Definitive (after finished)." (Vol. #4)

Recommendations were evidently made and followed, for in the proving dated Aug. 4th, considerable improvement is noted, by the letter of Hodgkins to Cuyler.

"By your direction I proceed this day to prove 3 gun barrels forged at the Arsenal, also the Colts pattern pistol barrel furnished by Messrs Griswold & Gunnison. The gun barrels stood all the tests we gave them up to 465 grains with 2 wads and 2 balls. 232 grains is all that is required. The pistol barrel also stood every test we gave it from 55 grains to 165 grains (the latter filled the barrel to within about ¾ inch of muzzle) with 2 balls well packed. Believing it impossible to burst either of the above by fair means, we deemed the above sufficient." (Vol. #4)

### FIRST REVOLVERS

Although some "100 revolvers were underway" as far back as July 1862 the finished product was not forthcoming until Oct. 14th. On this date, Lt. R. Milton Cary, later in charge of the C. S. Arsenal, Bellona, Va., near Midlothian, Va., but then Inspecting Officer, Artillery & Ordnance, wrote Cuyler:

"I have the honor to report that in obedience to your verbal order of the 12th, I on that day, and on yesterday, inspected twenty-two revolving pistols mfged after the model of the Colts Navy Pistol and submitted by Messrs Griswold and Gunnison.

Each pistol was taken apart and each part carefully inspected. They were then subjected to the following powder proof. Each barrel was fired separately with 54 grains of powder and 2 bullets.

Such as stood this test were then adjusted to the cylinders and the pistols fired, some with 1 bullet and some with 2 bullets, the cylinders being charged to their fullest capacity.

The barrels of 3 of the pistols bursted. One was found deficient because of a defect in the casting of the base. Another because of a broken hand spring. Another for a bursted tube or cone, and another because the ramrod catch was broken off.

The deficiencies of the last mentioned three will be repaired so as to pass inspection.

I therefore report that 18 out of the 22 pistols inspected by your order, as above, have passed inspection.

Numbers on pistols inspected; 3, 5, 9, 10, 11, 12, 15, 19, 20, 21, 22, 23, 24, 25, 26, 27, 28, 29, 30 (1), 30 (2), 31, 32. [Note duplication of serial #30—Editor]
Numbers 18, 20, 22, and 32 rejected."

Notice is made in the gaps between the serials up to #19, indicating considerable trouble at first and probably a large number of discards. From #19 on, the numbers run consecutively showing the manufactory was getting into more efficient operation.

Oct. 22, 1862, another letter from Lt. Cary to Major R. M. Cuyler, C.S.A., Comdg. Arsenal, Macon, Ga.

"On Monday last, the 20th, Messrs Griswold & Gunnison submitted for inspection, 22 of their pistols, numbers from #30 to #54 both inclusive. I have completed the inspection with the following results. Nos. 36, 39, 41, 48, & 52 were found to be fit for service.

The cylinder of #40 burst in proof.

The remainder were rejected for various defects in their parts such as want of temper in hand springs and bolts—a broken hand spring in proof—ratchet too short—main spring too short so as to be displaced by the recoil on discharge of pistol, or an improperly secured basepin.

The proof (powder) was the same hither to fore applied to a similar lot by your order." (Vol. #6)

The Griswoldville plant was desirably located, being on a railroad, and deep in Southern territory, and so free of enemy raids which so often interrupted competitive operations. But like all other Confederate enterprises, Griswold & Gunnison were no exception, and were plagued by lack of material.

### IRON FOR REVOLVERS

In the letter from Supt. Hodgkins to the then Captain Cuyler, dated July 16, 1862, already cited, Hodgkins verbally shook his head at Griswold & Gunnison's extensive use of iron for their revolvers rather than steel.

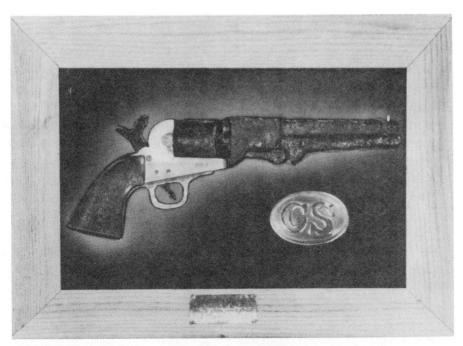

A Reconstructed Rebel

The following letter dated Dec. 1, 1863, from now Lt. Col. Cuyler to Major W. R. Hunt, of the Nitre & Mining Bureau, seems to vindicate the far sighted Sam Griswold. He was lucky if he even got iron.

"The forgoing bill of iron is urgently needed at this Arsenal to supply Messrs Griswold & Gunnison, contractors for making Navy revolvers. When the iron is furnished please let me know cost price?
From Shelby County Iron Works, Shelby County, Alabama.
7,000 lbs 2 inch round iron
11,000 lbs 1 inch square iron
2,000 lbs 1½ inch square iron
3,000 lbs 1¾ X 7/16 iron
1,200 lbs ¾ X ½ iron
1,800 lbs ¾ X ⅜ iron
2,000 lbs 1 X ½ iron
600 lbs ⅝ X ¼ iron
400 lbs ⅝ X 3/16 iron
1,000 lbs ½ inch square iron
4,000 lbs ½ inch round iron
5,000 lbs 5 inch round iron."

The fact was however that even iron was not plentiful. To cite an example. On Aug. 15, 1862, Cuyler wired Gorgas as follows:

............................................"Will you give Messrs Griswold & Gunnison, Contractors for Colts Revolvers, an order on C. T. McRae,

Selma, for iron?" (Vol. #101)

.......................... It was Sept. 4th before McRae, Agent, could reply,

"Capt. Absence for 3 weeks from this place has been the cause of my not answering your letters of the 16th and 18th ultimo.
I have today forwarded to the Shelby Co. Iron Works your order for the wrought iron for Messrs Griswold & Gunnison, and will ship you 8 tons (a carload) of pig metal by the 1st boat after tomorrow.
The Shelby Iron Co., is putting up a new furnace to run with cold blasts. It will be done in about 60 days, when they will turn out 7 tons #1 cold blast charcoal iron per day." (Vol. #6)

A receipt dated Sept. 29th, shows Cuyler paid $327.60 for a carload of pig iron whose weight was 16,380 pounds, and $22.60 for railroad freight from Selma. (Vol. 6)

The difficulty in transportation was not a minor item either. Cuyler's standard form letter for coal, coke, or iron from Selma was as follows:

................................................................ "I have sent in charge of a special messenger (..) crates to be filled with (coal, Coke, or Iron) in Selma, Ala. The crate will be loaded in platform cars, four on each, and are mounted on wheels so that they may be hauled across (Montgomery or Columbus) and put on board a steamer for Selma.
In this way I will avoid breaking bulk at Montgomery and Columbus. Should the agent have any difficulty in getting cars on which to load the crates at (Montgomery—Columbus), I beg that you will assist him to get them. It is a matter of utmost importance to have this (coal, coke or iron) at this Arsenal." (Vol. 101, Jan. 6, 1864)

## BRASS FOR REVOLVERS

Iron was not the only metal that was hard to come by. Upon first reaching Macon, Cuyler in urgent need of brass requested permission from Richmond to appeal to the patriotism of the local churches to "loan" their church bells for arsenal purposes. On May 7th he was told by Gorgas,

............................................................. "Capt. Your letter of the 30th ulto. has been received and contents noted. Get your works at Macon into operation as soon as possible, and get up 12, 4 gun batteries complete. Bells may be called for to be replaced at the close of the war." (Vol. 36)

Accordingly, the local churches were solicited to donate their bells to the "Cause." The results were gratifying. To quote a few. Cuyler from J. E. Evans, Pastor, Methodist Church, May 12th;

.................................... "This is to inform you that by order of the Methodist E. Church Mulberry St., of this city, their church bell is hereby placed at the control of the War Dept. of the Confederate States of America. It will be delivered when called for by a duly authorized agent."

Again, Cuyler from J. L. Jones, of Christ Church, Macon, May 26th; "At a meeting of the wardens and vestry of Christ Church held this P.M., it was

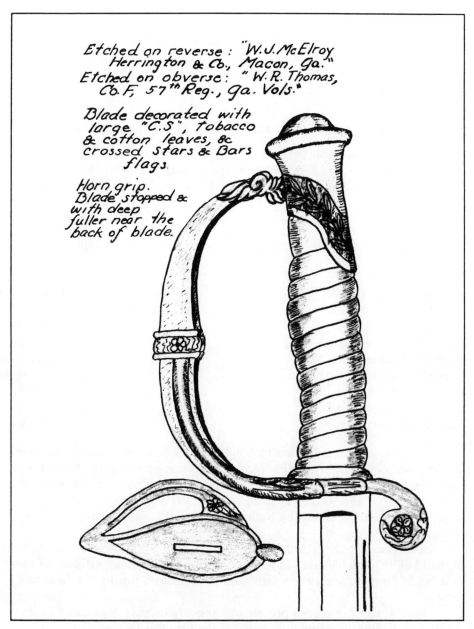

Confederate Sword made by W. J. McElroy & Co., Macon, Ga. *(in the Battle Abbey Collection)*

unanimously resolved to tender to you for the use of the Govt., their church bell. It is therefore at your disposal whenever you consider the exigencies of the country requires it."

It is no wonder that Woodrow Wilson in speaking of the Confederacy said;

"No cause, not of a religious nature had ever been so passionately or wholeheartedly embraced by a people."

These bells were gladly accepted by the Arsenal, and were melted down to make cannon, sword and bayonet hilts, and undoubtedly a portion of them went to supply Griswold & Gunnison. Those fortunate enough to own one of these revolvers may have a portion of the above named church bells in their possession.

## NUMBER OF REVOLVERS MADE BY GRISWOLD & GUNNISON

During the three year period that Griswold & Gunnison were in operation (Oct. 1862 to Nov. 1864) they averaged turning out about 100 pistols per month, for which the Govt. paid $40.00 at first, later raising this price to $50.00.

Vol. #33, gives us the following:

"estimate of funds needed for purchase of Navy revolvers" at the Macon Arsenal:

| | |
|---|---:|
| Needed for April 1863 | $ 6,000 |
| Needed for May 1863 | 6,000 |
| Needed for June 1863 | 6,000 |
| Needed for July 1863 | 6,000 |
| Needed for August 1863 | 10,000 |
| Needed for September 1863 | 12,000 |
| Needed for October 1863 | 12,000 |
| Needed for November 1863 | 10,000 |
| Needed for December 1863 | 10,000 |
| Needed for January 1864 | 10,000 |

This total estimate divided by $50.00 for each revolver would indicate 1,760 revolvers for the year 1863. During this year, however, the Macon Arsenal was also purchasing the .36 calibre (Navy) revolvers made by Spiller & Burr in Atlanta, and so the rough figure of 1,760 revolvers probably would apply to the total produced by both Spiller & Burr, and Griswold & Gunnison during this period.

As we know Spiller & Burr's output during this period was in the neighborhood of 760, the balance (1,000) is probably a pretty close estimate of the number of guns turned out by Griswold & Gunnison during the 10 month period.

This 100 gun per month production appears to have been more or less steady throughout the entire 3 year period the firm was in operation.

## DESTRUCTION OF GRISWOLDVILLE

Nov. 1, 1864 found Atlanta occupied by Gen. Sherman. Nov. 1, 1864, also saw Sam Griswold exhibiting the same farsightedness which to this point had brought him success.

Sam saw the "shadow of coming events" and in so seeing, took his pen in hand, and wrote Col. Burton of the Macon Armory to the effect that he no longer saw the benefits in private ownership of his pistol factory, and he preferred the C. S. Government would either rent or lease from him, his "machinery and negro workmen etc., for the manufactory of pistols."

Nov. 1st, Burton wrote Gorgas, Chief of Ordnance, enclosing the "letter from Mr. Samuel Griswold." Burton was against the Government taking over this factory, but proposed

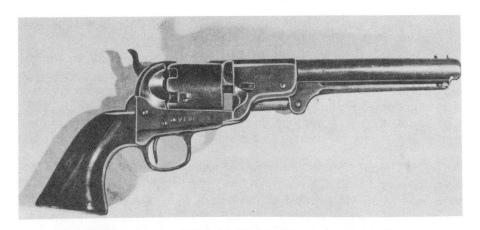

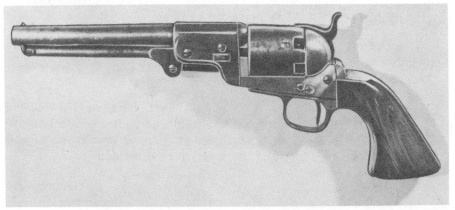

Right and left side of Griswold & Gunnison #2198 *(from the collection of the author)*

..................... "To raise the prices paid to Griswold to justify his continuing of the work.—I know the price paid hithertofore for pistols made at these works is much under that paid to other contractors." (Vol. 29)

Nov. 11, 1864, Gen. W. Tecumseh Sherman, U. S. A., began his easy but destructive march from Atlanta to the sea. His columns were spread to a width of 40 miles! The effect was very much like that of a railroad train coming in sideways. Damage done by those wanton raiders who destroyed for the sheer love of malicious destruction is beyond calculation, and even after 90 years the effects remain, and will remain—physically and spiritually.

A small division of the Confederate Gen. Joe Wheeler's cavalry pegged away on the advancing columns, and at Macon, Ga., Gen. Howl. Cobb brought out a division of Georgia Home Guards and Militia in a fruitless attempt to cut the huge serpent in two. Included in this militia were our old friends the "Rigdon Guards" of Augusta, Ga.

The serpent was not cut, but it was diverted and Macon was spared the torch.

State Militia (composed of workers, old men, and boys) against trained U.S. soldiers! The Battle of Griswoldville Station, was fought on a little knoll, about a mile east of the town. The dead were buried on the hillside and were never disturbed after that, and are said to remain there to this day in unmarked graves. The fighting was fierce, the casualties heavy on both sides, although there never was any doubt of the ultimate result.

Yankee reports of the battle remind me of an occasion during the last World War. It was twilight when a single Jap plane flew over a grouping of two or three hundred U. S. vessels of all types anchored in a Pacific atoll. Every ship opened fire and down went the Jap in flames. The following morning I doubt there was even one ship in that grouping which did not have another Jap flag freshly painted on her stack!

In the Official Records of the War of Rebellion we find "The pistol factory at Griswold Station was destroyed by the 10th Ohio Cavalry." (XLIV, page #54.) Col. Murray, 3rd Union Kentucky Cavalry reports, "We destroyed a pistol factory at Griswoldville." (Vol. XLIV, page 368.) Gen. Kilpatrick says; "The pistol factory at Griswoldville that we destroyed was very large and valuable." (Vol. XLIV, page 508.) and etc.

There seems to be no question but that the pistol factory was pretty thoroughly destroyed regardless of who did the actual job.

A newspaper account of the time states;

................................... "Every house in Griswoldville was burned by the enemy except Mr. Griswold's house, the residence occupied by Col. Grier, a few negro houses bordering on the branch and a small frame building occupied by one of the operatives of the mill."

One wonders how these escaped Sherman's wrath; certainly not through compassion?

The Battle of Griswoldville Station was fought Nov. 20, 1864, and this date marks the end of all pistols manufactured there.

However, on Feb. 21, 1865 (3 months later) Col. Gorgas, Chief of C. S. Ordnance, wrote Col. J. H. Burton of the Macon Armory, placing him in charge of various Confederate armories, and private contractors. Included in this list is the name of the "Griswold & Gunnison" contract revolver plant. Remember, however, communications were slow in those days. Evidently Gorgas was not aware that the factory had been destroyed beyond hopes of resurrection.

A month later, or on March 21, 1865, we find a letter from Burton to Col. Cuyler, Comdg. C. S. Armory, Macon, Ga., advising Cuyler that "the pistol manufactory of Mr. A. W. Gunnison" is being placed under his supervision by the "Chief of Ordnance" (Gorgas). Burton is "sure you will cooperate with me in getting Gunnison started again." (Vol. 29)

Same reference, March 29, 1865. Burton again comments to Col. Cuyler on the possibility of restarting Griswold & Gunnison, and this is the last official reference to be made on the Brass-framed Colt!

As this was written less than two weeks before Gen. Robert Lee's surrender at Appomattox, Va., there can be no question that operations were never resumed at Griswoldville, and that their last revolver was turned out prior to Nov. 20, 1864.

Before we turn the page though, one last letter from Burton, written March 31,1865 to "Messrs. Rigdon & Ansley, Pistol Mfgrs., Augusta, Ga." Burton notes that

............... "your establishment has been placed under my supervision by Col. Gorgas, making me responsible for the amount of work produced by your firm."

Burton wants to know if Rigdon-Ansley

............................................. "are now actively employed, and if so, what is the number of pistols turned out weekly or monthly, and what assistance do you require to produce maximum results?"

We have the advantage of time over Burton, and we know although he did not that the operations of Rigdon & Ansley, were as dead as those of Griswold & Gunnison. We also know from his record that Col. Burton was no fool, and must have seen the "handwriting on the wall" as clearly as did old Sam Griswold.

One can not help but feel an intense admiration for a man who can write such letters so close to the known end.

## MARKINGS AND THEIR MEANINGS

Having heard something of their background, let us take a look at the revolvers themselves. (see photo)

Measurements of these guns have already been given, and note has been made that except for the round barrel housing found in the early serials, all are pretty much the same. The markings also more or less conform to standard. The word "all" will therefore be used in describing these markings, although in the list of serials which follow will be found exceptions from which the reader may draw his own conclusions. All are stamped with the full serial number on cylinder, barrel frame, and lock frame. Some are stamped on the right side, others the left. This serial plainly applies to the total number of revolvers made up to that particular time of stamping.

The guns also contain a sub-serial number which is usually the last, last two, or first and last numbers of the primary serial. For example, if a gun's serial were #3421, the sub, or secondary serial will be either #1, #21, or #31. This secondary number is found on all parts—underside of trigger guard, backstrap, loading lever, cylinder pin, barrel wedge, even on the hammer and hand. It will not be found on the parts already stamped with the initial serial #3421.

This secondary number can have but one meaning. The three main parts were first put together—barrel, frame and cylinder. To these main parts were fitted the balance of the smaller parts, each being stamped with a portion of the initial serial to show that it definitely belonged to that particular gun.

Close examination reveals that all brass parts contain a type of "benchmark," in the form of a Roman numeral. This is cut, or chiseled into the brass roughly with a sharp pointed tool. It is not stamped. This numeral can be found on the side of the back strap, and on the underside of the trigger guard and frame. As it appears only on brass parts it obviously applies to some method of brass casting. The writer has never found this numeral to be higher than XX (20). It is believed this marking applies in some way to the number of brass castings which had been made from some particular mould. Possibly frame, trigger guard, and backstrap (for any one gun) were all cast at the same time in different but corresponding moulds. Each casting was marked with a numeral. After twenty castings a new mould was made and the process repeated.

There is still another mark which has been found on all guns, and it has been this mark, which has positively identified a gun in at least one instance in a piece otherwise devoid of markings which was declared by many who should have known better to be a "fake." This is a small cryptic mark which is stamped on all major parts—the barrel, on the underside near the loading lever lug; the side of the back strap; the underside or rear of the frame; the upper part of the trigger guard where it joins the frame and on the cylinder between the nipples near the safety pin. ALL (and here I use "all" literally) are found to contain this cryptic mark throughout. There is no mixing of this mark as is sometimes found to be the case with the numerical serials.

The term "cryptic" is used to describe this stamp because while it has been found in some cases to be a plain large letter "A," more usually it takes the form of a large letter backwards. One case it is noted as "S (on its side) AM," and in another case is found to be a series of four dashes which retain the same geometric pattern on each part.

This "cryptic" mark has been a source of great conjecture. It was first concluded to be the C. S. Ordnance Inspector's stamp, but if so, why would it not appear on all parts of other C. S. weapons?

The assembly line of today was unknown in the 1860's. Men worked alone or in small groups. We know from other revolver manufactories (as will be seen later on), that if 50 men were working in a factory, all fifty did not work on the same arm until it was completed. On the contrary, first a number of rough barrels, frames and cylinders were made. After these "roughs" were completed the men were broken into several goups or teams. One man was placed in charge of each team, he being held responsible in full for the revolver on which his team was working.

These "roughs," i. e., barrel, frame and cylinder, were drawn from some central storeroom in a semi-finished condition. After they were completely finished by the "team," they were laid on either right or left side and the serial #3421 was stamped on each of the three parts. As the balance of the parts were fitted to that particular gun they were stamped with a portion of the serial. The full serial #3421 was too large to stamp on the smaller parts so the secondary serial #21 was used and followed through on all remaining parts. After the gun was completed, it was taken down and the foreman inspected it closely—every part. He then placed his own stamp or "crypticmark" upon each major part, for remember, he was responsible for the proper working of gun #3421.

Many years ago when the "brass-frames" were supposed to have been made all over the South, collectors believed the cryptic mark "A" to have denoted the gun having been made at the Augusta Armory, the "R-backwards" showed it came from Richmond, the "F-backwards" that it came from Fayetteville, and etc. The logic of the above is not bad, and naturally supported the other erroneous theory that these guns were all made at different points.

To bear out the idea presently advanced, however, is a portion of a personal letter from Edwin Pugsley, on the letter-head of the Winchester Arms Co., dated Jan. 11, 1934.

"—answering your question as to what the letters stamped on the barrels and frames of the Confederate Colts might be, would say that I have no guess. It is conventional in more or less modern arms plants to stamp a code letter on a gun signifying the man who assembled that particular arm. This is a practice that has been used at Winchester for many years from time to time and I know that it has happened in other plants. Records are kept of these various letters in case of complaints about the operation of the arm.—"

## SERIALS OF GUNS KNOWN TO THE AUTHOR

Unfortunately the following list of serial numbers on Griswold & Gunnison revolvers is not as complete as it should be; the habit of acquiring full and complete data was not started soon enough. Even so it should be of some value to the collector.

| SERIAL NUMBER | SECONDARY MARK | ROMAN | CRYPTIC | SIDE ON WHICH SERIAL IS STAMPED. |
|---|---|---|---|---|
| None | None | Not known | R-backwards | |
| None | None | Not known | "AA" | |
| 8 | NK | Not known | NK | Not known |
| 19 | NK | Not known | NK | Not known |
| 45 | NK | Not known | NK | Not known |
| 46 | NK | Not known | NK | Not known |
| 70 | NK | Not known | NK | Not known (stamped C.S.) |
| 105 | NK | Not known | NK | Not known |
| 117 | NK | Not known | NK | Not known |
| 174 | NK | Not known | "T" | Not known |
| 262 | NK | Not known | NK | NK-(barrel frame cut away to aid loading). |
| 307 | NK | NK | NK | NK |
| 365 | NK | NK | NK | NK |
| 414 | NK | NK | NK | NK |
| 752 | 2 | NK | L-backwards | NK |
| 817 | NK | NK | NK | NK (stamped C.S.) |
| 906 | NK | NK | NK | NK |
| 949 | 19 | NK | NK | NK |
| 973 | 73 | II | "R"-backwards | NK |
| 993 | 3 | NK | NK | NK |
| 1101 | 1 | NK | "11-11" | Right - has Colt cylinder |
| 1144 | —cylinder only found on site of Griswold factory. | | | |
| 1146 | NK | NK | NK | NK |
| 1321 | NK | NK | NK | NK |
| 1369 | (my information on this piece is that cylinder is not original) | | | |
| 1415 | NK | NK | "BB" | NK |
| 1471 | NK | NK | NK | NK |
| 1472 | NK | NK | NK | N |
| 1474 | NK | NK | NK | NK |
| 1477 | 7 | NK | "L"-upsidedown | right |
| 1493 | 23 | NK | NK | NK |
| 1505 | 35 | NK | "D" | NK |
| 1510 | NK | NK | NK | right |
| 1514 | NK | NK | NK | right - barrel top frame round |
| 1554 | 24 | NK | "R"-backwards | serial stamped under frame. |
| 1564 | NK | NK | NK | NK |
| 1576 | 16 | II | "J J"-backwards | right |
| 1720 | NK | NK | "BB" | NK |
| 1750 | 10 | NK | "K" | left |
| 1802 | 10 | NK | "M" | NK |
| 1822 | 22 | NK | NK | left |
| 1840 | 40 | XIII | S-backwards | NK |
| 1852 | 52 | NK | H | left |

| SERIAL NUMBER | SECONDARY NUMBER | ROMAN | CRYPTIC | SIDE ON WHICH SERIAL IS STAMPED |
|---|---|---|---|---|
| 1862 | NK | NK | NK | NK |
| 1863 | 3 | NK | "K" | right |
| 1915 | 55 | NK | NK | NK |
| 1934 | NK | NK | "H" | NK |
| 1978 | NK | II | NK | right |
| 2040 | NK | NK | NK | right |
| 2044 | 4 | III | "N" | NK |
| 2093 | NK | NK | NK | NK (stamped CS) |
| 2096 | NK | NK | NK | NK (stamped CSA on frame) |
| 2116 | NK | NK | NK | NK |
| 2145 | NK | NK | NK | NK |
| 2184 | 24 | XVII | "R"-backwards | NK |
| 2198 | 8 | XI | "A" | right (this gun belonged to Major Brown of the 7th Va. Infantry, and is so marked on backstrap). |
| 2231 | 41 | NK | "N" | right |
| 2246 | 56 | NK | S-on side AM | NK |
| 2282 | NK | NK | NK | NK (barrel cut to 6 inches). |
| 2363 | NK | NK | NK | NK |
| 2365 | NK | NK | "O" | NK |
| 2389 | 6 | NK | "A" | NK |
| 2419 | 49 | VI | "II II" | left |
| 2437 | 7 | NK | "T" | NK |
| 2457 | 27 | NK | "CC" | NK |
| 2470 | NK | NK | NK | NK |
| 2651 | NK | NK | "BB" | left (cylinder not original). |
| 2695 | 25 | IX | F (backwards) | right |
| 2696 | NK | NK | NK | NK |
| 2708 | 8 | NK | "11" | NK - has re-enforced frame. |
| 2763 | NK | NK | NK | NK |
| 2855 | NK | NK | NK | NK |
| 2860 | NK | NK | NK | NK |
| 2922 | NK | NK | NK | right |
| 2987 | 17 | NK | NK | NK |
| 3085 | 25 | NK | "L"-backwards | NK |
| 3106 | NK | NK | NK | NK |
| 3334 | NK | NK | NK | NK |
| 3399 | 9 | NK | "DD" | right |
| 3424 | NK | NK | NK | NK |
| 3447 | 7 | NK | NK | right |
| 3606 | 44 or 45 | NK | "J" | NK |

Certainly if the above listing were more complete some sort of a pattern would appear, but even so, examination will disclose some facts. We note that at least two of the revolvers were made without any serial numbers at all. If two have survived at this late date, we can but wonder how many might have originally been made without serials and why they were so made? Were they the first, or the last? Further note that the only two revolvers within close range of each other on which complete markings have been obtained are #2184, whose Roman numeral is XVII, and #2198 whose Roman numeral is XI. This would in some way substantuate the theory that the Roman numerals went only to XX (20) and then started again, for if #2184 is XVII, it would follow #2185 was XVIII, #2186 would be XIX, etc., until we reach #2198 which would be, and is XI.

A variation is noted in #2231, which has as a secondary serial #41, rather than the expected #31, or 10 higher than is normally found. Looking at #2246 which is a nearby serial, we see its secondary to be #56, or still 10 higher than is normally found. Is this merely coincidence, or is #2247, marked with the secondary #57, or #47? What is the reason for this variation?

Because of the lack of records, almost all studies on Confederate arms seem to end in a question mark, and this present work is no exception.

The writer is well aware of the gaps which appear in this monograph, main one of which is the total lack of any information as to A. W. Gunnison, partner in the firm of Griswold & Gunnison.

Diligent search through all possible available sources gives no clue as to whom he was, whence he came, or where departed; or indeed even his actual connection with the revolver factory. One might make the guess that possibly he had been a foreman at the Griswold Gin Factory, elevated to partnership level with the advent of revolver making. This theory does not seem so good when it is found that his name is totally unfamiliar to persons who have resided in and around Griswoldville for over 80 years.

It seems odd that a person who was a partner in a firm which manufactured at least 3606 revolvers (almost as many as all the other Confederate makers put together) could disappear leaving no trace of his previous background or postwar connections.

Possibly our inherent love of mystery accounts in part for the fascination which attends any study of Confederate Ordnance—so far we go and no farther.

THE END

Since originally preparing the manuscript on the Griswold & Gunnison revolvers, the following serial numbers have been supplied by my good friend, Edward N. Simmons.

| Serial Number | Secondary Serial | Roman Numeral | Cryptic Mark |
|---|---|---|---|
| 133 | 33 | V | not known |
| 135 | NK | NK | NK |
| 202 | NK | NK | NK |
| 301 | NK | NK | NK |
| 608 | 8 | 1111 | LL |
| 696 | NK | NK | NK |
| 763 | NK | NK | NK |
| 776 | NK | NK | NK |
| 1041 | NK | NK | NK |
| 1209 | NK | NK | NK |
| 1486 | NK | NK | NK |
| 1516 | NK | NK | NK |
| 1550 | NK | NK | NK |
| 1635 | NK | NK | NK |
| 1782 | 12 | V | X |
| 2110 | NK | NK | NK |
| 2127 | NK | NK | NK |
| 2273 | NK | NK | NK |
| 2394 | 24 | 1111 | J (backwards |
| 2614 | NK | NK | NK |
| 2909 | NK | NK | NK |
| 2941 | NK | NK | NK |
| 3094 | NK | NK | NK |
| 3193 | NK | NK | NK |
| 3235 | NK | NK | NK |
| 3355 | NK | NK | NK |

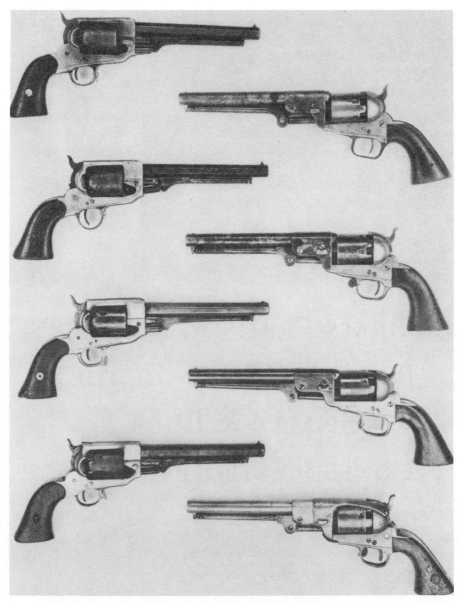

Group of Spiller & Burrs and Griswold & Gunnisons
The 4 Spiller & Burrs are: #128, #131, #150, #214
The 4 Griswold & Gunnisons are: #949, #1750, #2419, #2708
*(from the collection of Harry Brooks)*

# PART II

## The Brass-Framed "Whitney's," Being the Story of the Spiller & Burr Confederate Revolver

## CONFEDERATE "WHITNEY'S"

Confederate "Whitney" revolvers, 6 shot, .36 calibre, were made under the firm name of Spiller & Burr. Some are so marked on the top of the octagon barrel. Others bear only the serial number on major parts, and "C. S." stamped on the right or left side of their all brass frames. Still others are totally without markings.

Although entirely "legitimate," the parentage of these revolvers is somewhat unusual in that they were conceived prior to the Civil War by one Samuel C. Robinson of Richmond, Va., but when actually "born" in the spring of 1863, their "father" was the firm of Spiller & Burr.

Later on, we will explore Samuel C. Robinson's part in this venture, but first we will concern ourselves only with those persons who were connected with the enterprise at the time the revolvers were forthcoming.

Those persons were: Edward N. Spiller, David J. Burr, and James Henry Burton.

## EDWARD N. SPILLER

A letter dated Feb. 10, 1920, from J. P. Smith of Raphine, Va. (a former trooper under Col. Mosby, CSA) tells us something of the background of Spiller. Says Mr. Smith;

...................."—I know nothing of Burr as a partner (in the revolver factory), but Edward N. Spiller was an uncle of mine. Before the war he ran a commission business in Baltimore, Md., but being a strong Southern man (sympathizer), and overly high-strung, had to leave there and come South. He lived at my father's house in Richmond for some time until he went to Atlanta. On one of his visits back from Atlanta, he brought one of his pistols, and gave it to my father. I remember it perfectly although I was in the Army at the time. Edward N. Spiller lived in Baltimore for years, before and after the war. His only daughter married a prominent iron manufacturer of near there, but I fail to remember his name."

## DAVID J. BURR

David J. Burr, a native of Richmond, Va., had long been engaged in building locomotives. The company bearing his name was located at 5th and Byrd Sts., and in Jan. 1836 built a steam engine for the Richmond, Fredericksburg & Potomac Railway. In 1842, Burr, Poe & Sampson built the "Governor McDowell," a 90 by 13 foot steam packet, the first and only one to appear on the James River and Kanawha Canal, terminus of which was in Richmond, Va. In 1852, was established the Burr & Ettinger Locomotive Works of which Burr was an active member. He was a man of some prominence, and although his connection with Spiller & Burr appears more financial than active participation, there is no doubt that his name added considerable prestige to the firm.

## JAMES HENRY BURTON

At the time of the Spiller & Burr venture, Burton was a Lieutenant Colonel in the Confederate Army, assigned to the Bureau of Ordnance, and held the official title of Superintendent of Armories, C.S.A. This was his full time job, but at the same time, he held others, anyone of which the average man would be content to consider fulltime. Lt. Col. Burton however was decidedly not the average man.

His life and innumerable connections would make an extremely interesting book in its own right, and there is no lack of available material for some future historian to do just that. This wealth of material has created in this work the problem of what to include, and what to leave out.

His obituary appeared in the "Baltimore Sun" of Oct. 19, 1894.

Says the "Sun":

"Winchester, Va.—Col. James Henry Burton, one of the most prominent citizens of this section, died at his home here this evening (Oct. 18) after a brief illness of pneumonia, having been sick less than a week. Col. Burton was born of English parents, Aug. 17, 1823, at Shennondale Springs, Jefferson Co., Va. After receiving an education at the Westchester Academy, Penn., he entered at the age of 16 a machine shop in Baltimore to learn the business of practical machinist, and graduated from there four years later.

In 1844 he took employment in the rifle works of the U.S. Armory at Harper's Ferry, and was appointed foreman in 1845. He afterwards received the appointment of Master Armorer, which he held until 1854. In 1855 he accepted the appointment of Chief Engineer of the Royal Small Arms Factory at Enfield, near London, England. Five years afterwards he returned to Virginia in consequence of failing health. In June 1861, Mr. Burton was commissioned lieutenant-colonel in the Ordnance Dept. of Virginia by Gov. John Letcher, and placed in charge of the Virginia State Armory with instructions to arrange for the removal thereto with the utmost dispatch the machinery, etc., captured at Harper's Ferry, and to place it in position for use. Within 90 days from the date of his commission he had the machinery at work in Richmond, producing rifles of the United States patterns.

The following Sept. he was commissioned by Pres. Jefferson Davis, Supt. of Arms, with the rank of Lieutenant-Colonel. During the summer of 1863, Col. Burton was ordered to Europe on business for the State Dept. At the close of the war, and after recovering from a severe illness, Col. Burton with his family spent over three years in England. Upon his return, he located in Loudoun Co., Va., where he resided until 1871, when he again went to England at the instance of a private firm in Leeds, to take direction of a contract entered into with the Russian Govt. for the supply of the entire plant of machinery for a small arms factory on a large scale at Tula, in Central Russia, for the manufacture of the Berdan rifle, and with the view of ultimately going to Tula as an officer of the Russian Govt. to take the technical direction of the factory. He was constrained to resign his position and returned to Virginia in the fall of 1873, since which time he has been following the peaceful pursuit of a farmer near Winchester."

Add to the above that Col. Burton found time to begat himself a family of 14 children, and it must be agreed that this was a man who lead the full life. Burton's seventh child, Frank, was the Master Mechanic at the Winchester Repeating Arms Co. for years.

### CONTRACT BETWEEN SPILLER, BURR AND BURTON

The fall of 1861 found the main actors of this story in Richmond, Va. All of them were highly imbued with the "war-fever." Apparently Spiller had fired Burr with the idea of pistol making, and the two realizing Burton's "know-how" plus his connections in the Confederate Ordnance Bur., sought his aid both in securing a Govt. contract, and for assistance in the actual operations. Nov. 20, 1861, saw a meeting of the minds of the three, and a contract was drawn up and signed as of that date.

Deleting the ponderous "whereas's etc.," this agreement provided:

#1—Burton was to secure from the C.S. Govt., a contract for 15,000 Navy size revolvers at $30.00 for the 1st 5,000, $27.00 for the 2nd, and $25.00 for the 3rd. Funds necessary for the starting of this venture was to be advanced by the Govt. without interest if the enterprise succeeded. Spiller & Burr were to give personal security for double the amount of money so advanced.

#2.—Burton was to superintend preparation of the necessary plans, machinery, tools; the erection of a necessary building; start the machinery, and superintend the manufacture of pistols after the manufactory was started; and was responsible for operation of all mechanical arrangements required to be provided for the purpose. He was however to give only so much of his time as was not required of him in filling the appointment at Supt. of the C.S. Armory which he then held.

#3.—Burton was to commence without delay and complete the order for 15,000 pistols in the time specified, which was to be no longer than 2 years, and 6 months from the date of the contract.

#4.—Spiller & Burr, for all of the above, were to pay Burton $2,500 as soon as the Govt. contract was procured, then a further sum of $2,500 upon the completion of the 1st 100 pistols. Still in addition, Burton was to receive 1/3 the profits for each year the business was in operation.

#5.—Burton was not to be held responsible for any debts, or obligations incurred by the firm, that he was not a partner, and his connection with the business was strictly as specified.

#6.—The contract was binding for a period of 3 years, or until Nov. 1864.

This contract is signed by Edw. N. Spiller, David J. Burr, and Jas. H. Burton.

Burton let no grass grow under his feet in obtaining the following contract from the Govt. This too has been skeletonized for easier reading.

## CONTRACT BETWEEN SPILLER & BURR, AND THE C.S. GOVT.

Section #1.
Article #1.—The War Dept. guaranteed an order for 15,000 revolvers, of a pattern substantially the same as that known as "Colts."

#2.—An advance of $20,000 immediately on conclusion of the contract, the further sum of $20,000 at the end of 3 months, another sum of $20,000 at the end of 6 months, to be made by the Govt. to the contractors who in turn were to give satisfactory personal security in the sum of $120,000. The money advanced to be free of interest, but such money would later be returned with an interest rate of 8% per annum. Should the $60,000 be found insufficient, further advances were to be made, providing the entire amount did not exceed $100,000.

#3.—The Govt. agreed to pay $30 for each of the 1st 5,000 pistols, $27.00 for the next 5,000, and $25.00 for the final 5,000.

#4.—The Govt. agreed to compensate the contractors for any loss sustained, provided such loss be the result of invasion by the North.

#5.—Custom duties will be remitted on any materials imported for the purpose of the contract.

#6.—Preference to be given contractors over all others in case it was desired to increase the orders for pistols.
#7.—The War Dept. to inspect on pistols presented within a 2 week time from the date of presentation.

On their part, Spiller & Burr agreed to:

Sect #2.

Article #1.—Erect and put in operation within the C.S.A., a factory capable of producing no less than 7,000 revolving pistols per year.
#2.—Supply the War Dept. as follows:—4,000 pistols by Dec. 1, 1862, 7,000 by Dec. 1, 1863, and 4,000 June 1, 1864, or at an earlier date if convenient.
#3.—Nothing but the best materials that can be obtained were to be used, but in the event of the impossibility of obtaining steel for cylinders and barrels, it is agreed that iron of suitable good quality be substituted. The lock frames may be of good tough brass if properly electro-plated with silver.
#4.—Pistols will conform to the model furnished by the War Dept.
#5.—All pistols finished will be submitted to the authorized agent of the War Dept. in accordance with special instructions defining the system of inspection and the tests to be applied in the examination of the finished arms.
#6.—Pistols will be presented in the City of Richmond in lots of no less than 100 at a time, unless otherwise desired by the War Dept.
#7.—Spare parts for the pistols will be furnished by the contractor at reasonable prices.
#8.—As a means of securing the mechanical success, it is with the consent of the War Dept., that James H. Burton will give such assistance etc., as will not interfere with his public duties and obligations.

This contract is dated Nov. 30, 1861, and shows many things, among which are the high hopes that could lead Burton, an experienced armorer, and two persons totally unacquainted with weapon making to the signing of a contract for 15,000 revolvers to be delivered within a 2½ year period! The contract also shows that even in 1861 the newly formed Confederacy was fearful as to a sufficient supply of steel. Still further it showed that considerable material would have to be imported, and that the Southern States were not self-sufficient so far as material of war was concerned. As a Confederate colonel said, his men needed only rifles, clothes, great coats, knapsacks, canteens—in short, everything but the will to fight. Of this last, there was always plenty to be found in the South.

## PUBLIC OPINION AGAINST BURTON

As well might be expected, there was considerable public opinion against Burton, who could hold public office and yet share in private enterprise which was of his own instigation, and further, started with public funds.

The matter came to a head in April 1862 with a direct attack upon Col. Burton by the Hon. J. B. Heiskell, M.C. of the House of Representatives on the floor of the Confederate States Congress.

Burton was a man of considerable spirit, and immediately resigned his appointment as Superintendent of Armories.

The following four letters from the principals in this little side-drama best speak for themselves.

## BURTON RESIGNS

Burton to Col. Gorgas, Chief of Ordnance, April 21, 1862.

"Colonel:

I hereby respectfully tender my resignation of my appointment as Superintendant of Armories to take effect as soon as you can appoint a successor to relieve me. Although zealously anxious to do all in my power towards providing the means of defense so necessary at the present moment to our beloved South, yet my sense of self respect will not permit me to retain office under the General Government for one moment longer than is imperatively necessary in the face of such a wanton attack upon my private and public character as was made upon the floor of Congress on yesterday, and which was suffered to pass unrepelled.

I claim to have been a faithful servant to the C. S. Government during my tenure of office under it and I think I have rendered efficient service and therefore had a right to expect better treatment at the hands of those charged with the legislation of the country. I trust therefore that you will be pleased to accept my resignation."

## COL. GORGAS COMES TO HIS FRIEND'S SUPPORT

Gorgas to Hon. J. B. Heiskell, M. C., April 21, 1862.

"Dear Sir:

I beg leave to call your attention unofficially to what you stated in reference to Col. Burton, Supt. Armory. The weight due to what has been told you against him may be judged of, from the character of one item viz; that which refers to his interest in a private contract. Mr. Burton was permitted at my request to give his services when not in conflict with his other duties to Messrs. Spiller & Burr, contractors for pistols. I asked for this (and in fact suggested the matter to Col. Burton) in order to secure the erection of good machinery for a pistol manufactory and this permission is inserted into the contract with Spiller & Burr, and forms a part of its terms. I do not hesitate to pronounce him the most competent and valuable officer we have connected with the Bureau, and you will find on inquiry that my high opinion of him is fully endorsed by the head of the Bureau of Ordnance of the Navy, Capt. Minor. I trust your sense of justice will induce you to take an early opportunity to undo the injury you have done him upon unreliable testimony. I have to add that he has offered today his resignation stating that his sense of self respect will not permit him to retain his position under such an attack. I shall of course use every effort to prevent its acceptance and to have justice done to him."

## HEISKELL RETREATS FROM HIS ORIGINAL STAND

Heiskell to Col. Gorgas, April 22, 1862.

"Dear Sir:

I regret very much that your communication came to hand too late for me to mention the facts you state to the House. The Speaker was just on the point of announcing the adjournment and it was done before I got through your letter. I was careful to state when the matter was before the House that I made no charge against Col. Burton when (sic) I did not even know I stated that I had been informed that certain things were so which in the latter stage of the session was impossible to enter into; that I felt it my duty to mention them as cautionary, but at the same time that I would caution the House not to place reliance on them to the injury of Col. Burton.

That we all knew the readiness with which such charges were made, especially against men of position and that I hoped none would place credence the statements as being facts. I did not consider that I was making any attack on Col. Burton and in this carefully stating, simply what I had heard, no one can regret more than I if he shall deprive the country of services really valuable from a mistaken and over sensitive view of what I had said.

I will say further lest any misapprehension should exist that I did not receive any of the information of which I spoke from the Armory officers."

### SEC. OF WAR RANDOLPH POURS OIL ON TROUBLED WATERS

Sec. of War, George W. Randolph, to Col. Burton, April 29, 1862.

"Sir:

I have received your tender of resignation in consequence of attacks made upon you in Congress which had been suffered to pass unrepelled. I respectfully decline to accept your resignation. Your qualifications and services are too well known and too highly appreciated by this Department to permit its acceptance upon such grounds. The attack referred to has been publicly disavowed and even had it not been, a resignation would not have exposed its injustice, but on the contrary, in the opinion of some, might have given it some weight. In justice to yourself therefore as well as for the sake of the Public service, I decline to accept your resignation."

### BURTON AND THE ALEXANDER CARBINE

In Dec. 1861, Col. Burton entered into much the same type of agreement with one Charles W. Alexander, as he had the month previous with Spiller & Burr. Alexander was the supposed inventor of a carbine which was later to develop into the "Richmond Sharps." In their agreement, Alexander was to avail himself of Burton's assistance with a view of introducing his carbine into the army. Burton was to use his influence with the Chief of Ordnance for said introduction, and for so doing, was to receive half interest in the patent. Instead, he received the following sharply worded rebuff from Gorgas himself;

"Sir:

Pressure of business has heretofore prevented the notice I intended to have taken of your proposed agreement with Mr. Alexander in reference to his carbine. It is by no means proper that you should make your influence with me enter into stipulations of a contract; a knowledge of this would destroy the confidence I have in you. The reservation that you are to devote only such portions of your time and attention as is compatible with your duties relieves you from any imputation of taking advantage of your position in that respect. You have in one instance, at my suggestion, been permitted to assist with your well known experience, a party who had a contract with this Bureau; this indulgence to you was given because it was likely to prove of real advantage to the Dept. in the end; the indulgence was an exception and must not be permitted to pass into a rule."

## BURTON AND THE McNEILL CARBINE

The Richmond "Daily Examiner" June 8, 1861, contains a notice of one Thomas E. McNeill, requesting contributions to the "C. S. Armory & Foundry Co." Capital $1,000,000.00 to fabricate artillery of all types, rifles, muskets, pistols, swords, bayonets, rockets, and all munitions of war, Macon, Ga., Thomas E. McNeill, Acting Supt., Glaze & Radcliffe, Agents. (This was the same William Glaze, operator of the Palmetto Armory, while Radcliffe's name is frequently found stamped on many imported English Tranter revolvers.)

The C. S. Armory & Foundry Co. was never organized beyond the paper stage, and we find no further mention of this concern that was to manufacture all weapons of war, but in April 1862, Mr. McNeill again comes into the picture in connection with a carbine he proposed to make.

April 25, 1862 finds an agreement between McNeill and Burton in which Burton was to "assist McNeill from time to time so far as my other obligations and duties will permit." For this, he was to receive $5,000 for certain of his own personal drawings, and would receive an additional $5,000 if the drawings were lost. He was also to receive so much for each carbine manufactured.

McNeill with Burton's assistance secured a contract from the Govt. for 20,000 such carbines at $45 per!

Both the Alexander and the McNeill carbine ventures were complete fiascos. No carbines were forthcoming therefrom.

In my files are the complete papers covering these two ventures, and in each instance Burton's first friendly letters later turn to bitter invective when things were not to his liking. One gathers that the Col. Burton, was a mite difficult man with whom to cross swords.

My admiration for Col. Burton is high, and without attempting in the slightest degree to detract from his war record; we can only state that there seems little wonder at public resentment against his private activities which were planned to secure him considerable sums of money at a time he was employed by the Govt. Let it be further stated, however, that there is no indication he defaulted from the strict letter of any contract into which he entered, and his energies must have been without limit.

## THE SPILLER AND BURR PLANT IN RICHMOND, VA.

Having actually secured a Confederate contract on Nov. 30, 1861, the firm of Spiller & Burr, immediately set out to make the 15,000 revolvers. Their initial step in this direction was the selection of suitable quarters.

It is at this point that S. C. Robinson comes into our story for a brief moment.

Searching through available records which pertain to Richmond, Va. in 1860-61, mention is found of a "Robinson Revolver factory" which "anticipated the manufacture of Whitney type revolvers."

Samuel C. Robinson was the owner of the factory in question. He was from a family long prominent in the social and financial affairs of Richmond, and during the peaceful ante-bellum days operated the Belvidere Planing Mills, which the 1859 city directory shows located at 9th and Arch Sts.

When war with the North became "possible," and the old Virginia Armory was reactivated in 1860, Robinson converted his planing mill into an armory for the purpose of modernizing the flint and steel Virginia Manufactory muskets, rifles, and pistols to percussion, while the Virginia Armory itself was busily engaged in retooling.

When war with the North became "probable," Robinson being in a position to know how lacking in arms was the South, proposed to rectify this deficiency in part by supplying the State of Virginia with revolvers at $18.00 per. His proposal stipulated revolvers of the Whitney model. Before his offer was acted upon, Robinson changed his price to $20.00. At this time (April 1861) war was no longer "possible" or even "probable" but "inevitable." Robinson's offer was accepted by the Virginia legislature just a few days before the State seceded. (or LI Part II, p. 48, April 27, 1861)

There is not the slightest indication that any of these weapons were ever produced, but obviously to have delved so deeply into this proposed manufacture Robinson must have drawn up plans, patterns, costs, and possibly even made some machinery. All the while the records reflect he was altering the old flint and steel arms with regularity.

This latter activity continued until the fall of 1861; meanwhile he was progressing (we assume) with his plans to manufacture his Whitney model revolvers, but meanwhile March 1862 became sidetracked in the manufacture of a carbine which closely follows the Sharp's model. These guns are known to the arms collecting fraternity as "Richmond Sharps" and from their inception until March 1863 when bought out by the Govt., were turned out by S. C. Robinson, who operated his establishment under the trade name of "S. C. Robinson Arms Manufactory." His carbines are so marked. After their purchase by the Govt., they are marked simply "Richmond, Va."

Taking into consideration that the first carbine was not actually turned out until Dec. 1862, the fact that Robinson managed to produce some 1,800 stand of arms before being bought out in March 1863, speaks well of his ability. He is one of the few who did what he said he would.

Despite the fact that the Spiller & Burr contract called for "revolvers substantually of the Colt's pattern," when finally forthcoming, they are found to be of the Whitney model. Remembering the initial Robinson contract that called for Whitneys, and that none were ever produced, it seems reasonable to suppose that Spiller & Burr assumed the Robinson contract, rented a portion of his armory, and bought what plans, patterns and machinery that had already been started.

To give some strength to "theory," we quote a portion of a letter from George W. Rogers, son of a former Robinson factory operative. Writes Mr. Rogers as of July 17, 1928.

"My father is too old to read, and so I read him your letter (requesting information as to S. C. Robinson). During the war my father was too young to enlist in the regular Confederate Army, but did active service in the Elliott Grays, or Home Guards. He says he is under the impression that the Samuel C. Robinson referred to in your letter was the proprietor of the Robinson Revolver Factory, and a Mr. Lester was superintendent. The factory was on the south bank of the canal (James River & Kanawha Canal), just east of the railroad bridge crossing the James River, and known as the Petersburg Bridge. This would make the location about 8th St., on the south side of the canal extending to the river. The revolver factory was moved the latter part of 1862 or 1863 to some point South."

All this fits in with what we know of Spiller & Burr.
Mr. Rogers continues;

........................"My father was employed for a while in the Robinson Revolver Factory which was also called Robinson's Carbine Works, and my mother was employed just a few blocks away at the Confederate Laboratory."

Whatever the set-up of Spiller & Burr in Richmond, things did not work out as they had originally hoped and May of 1862 found E. N. Spiller in Atlanta, Ga. looking for a proper site at that point.

This removal was probably undertaken because Col. Burton, their guiding light had been ordered South to establish a permanent C. S. Armory, and they wished to remain physically close to his establishment.

The firm had originally intended to operate under the name of the "Richmond Small Arms Manufactory," but as their first revolver was not turned out until after leaving this city, they reverted back to the name of their principals—Spiller & Burr.

Very definitely no revolvers were made while in Richmond, and their operations here must have been limited to getting together their machinery, etc.

Col. Burton kept notes and diaries, which although not always continuous, are nevertheless of great assistance in establishing dates. On May 15, 1862, he drew up an "Order of Operations on various parts of the pistol to be manufactured by Spiller & Burr." His use of the future tense "to be manufactured" certainly indicates that up to that point (May 15th) no revolvers had yet been manufactured, and as this "Order of Operations" is dated in Richmond, it shows the firm was still in that city.

## REMOVAL TO ATLANTA, GA.

On May 27, 1862, Lt. Col. Burton was relieved of command of the Richmond Armory by Capt. W. S. Downer, and according to his diary, he and his family left the following day for Atlanta, "to establish an extensive armory at, or near that place." He remained overnight in Lynchburg, Va., and resumed his journey at 5 A. M. next morning via the Bristol, Knoxville, Cleveland and Dalton Rwy. (now a portion of the Norfolk & Western Rwy.). He arrived in Atlanta on Saturday morning May 31st, at 2:30 A. M. and took rooms at the Trout House. Four days to cover 500 miles in those days was not bad time.

After a few hours rest, the untiring colonel spent the morning "walking over Atlanta with Mr. Spiller," showing that Spiller had already reached that city. That same afternoon, Burton "drove out to the residence of L. P. Grant and spent several hours with him discussing matters of interest in connection with the location of the proposed C. S. Armory."

This proposed Armory was planned as a long range, and permanent establishment, and not just a wartime measure, consequently, much care was expended as to its location, etc.

From the date of his arrival in Atlanta until June 11th, Burton put in many busy hours seeking just the right site for his armory. On June 11th, he writes Col. Gorgas;

"Col.:

I have the honor to hereby acknowledge the receipt of your instructions of the 4th & 5th inst. on the subject of the location of the proposed armory and the temporary erection of the stock machines at as early a day as practical. Since my arrival here, I have been actively engaged in examining the country immediately surrounding it with a view to the selection of a good site for the armory, but in consequence of the conditions necessary to be fulfilled, I have found it rather difficult to find a location affording the necessary facilities.

Three essential conditions must be fulfilled viz: a sufficient level tract for the erection of the necessary buildings; 2nd, a supply of water from a running stream for steam and other purposes; and lastly, proximity to a railroad and means of connection therewith by a siding.

The country around Atlanta is so broken and rolling, that it is almost impossible to find an acre of perfectly level ground, but I have succeeded in finding a location which—whilst it is the only one that would answer the purpose, is an admirable one—high and commanding—is bounded on one side by the line of the Macon & Western Rwy., and is just outside the city limits, and a good supply of water can be obtained from a stream of spring water two or three hundred yards distant. The tract was formerly used as a race ground, and is one mile in circumference, containing 40 to 50 acres. It is now rented by Capt. M. H. Wright, and he is erecting a number of temporary buildings on it for laboratory purposes. The owner resides in Barnwell, S. C., and a few weeks ago, stated to Capt. Wright that he would sell the tract for $15,000. I have conferred with that officer on the subject and we have decided to send a trusty person to treat with the owner of this tract with a view to its purchase by the C. S. Govt., and he will leave tomorrow morning.

It will be necessary to purchase in addition, a few acres of ground adjoining in order to secure the water supply. As this is the only parcel of ground in this vicinity that will answer the desired purpose, I respectfully request you consent to its purchase on the best terms I can secure, and I shall be glad to receive your answer by telegraph, as property changes hands so rapidly here now that unless the purchase be made promptly, the chance may be lost, which I should regret.

Capt. Wright agrees with me in the favorable opinion I have formed of the location referred to. If you agree to the purchase, I will have the tract carefully surveyed and a plan made of it, a copy of which I will forward to you.

I have visited Macon, and although that city affords some advantages in point of favorable ground, and is also a more desirable place to live in than Atlanta, yet it has the disadvantages of being much warmer and farther removed from the coal regions. Upon enquiry, I find it will cost at least $2.00 per ton more for coal in Macon than in Atlanta, in consequence of the transportation.

Building would be cheaper in Macon than in Atlanta, as the lumber supply is derived from that vicinity. All things considered, I have concluded that *this* is the best point for the location of the Armory, apart from the question of safety. Macon is certainly a safer point than Atlanta, but at present there would seem to be no doubt as regards the safety of the latter place, notwithstanding the late attack upon Chattanooga, from which point the enemy have now retired without having accomplished anything of value to them.

Chattanooga is a most important point to hold, for without it, the whole coal supply would be cut off in this direction. This was the case last week during the attack I refer to, but traffic has been again resumed.

Stores to an immense value are now stored here and no doubt the enemy would be glad to reach this point in order to capture or destroy them if possible. There might be a possibility of his doing so if he held Chattanooga, which however I trust he will be prevented from doing. It is not improbable that he may make another demonstration before long upon that point.

With reference to the erection of the stock machines—I find but one shop in this vicinity that will answer the purpose, and which can be secured by purchase at the price of $20,000 together with all the machines and fixtures it contains, pertaining to the manufacture of doors, sash and blinds etc. It is the same property to which your attention was called on or about 5th of April last by Capt. Wright, and is owned by Messrs. Pitts & Cook.

You will doubtless find upon file in your office a letter from Capt. Wright on this subject accompanied by a plan of the premises and a description of the property together with a list of machines etc., proposed to be sold with it.

The various machines for working wood will be very useful to either Col. Rains, or to Capt. Wright in connection with the manufacture of gun carriages, etc. There is a pretty good 30 HP engine and boiler now running on the premises and shafting and pulleys that will answer a good purpose for driving the stock machines.

About 70 feet of the length of the main building (all buildings are of frame and cheap construction) will be sufficient for the accommodation of the stock machines and the whole of the balance (100 feet) I propose to use as a machine shop in which case I shall be able to start a good force in a short time on the unfinished machinery to arrive here: all of which well be very desirable to do.

Unless these premises are secured, it will be impossible to erect buildings & engines etc., in time to keep up the supply of stocks to the Richmond Armory, and as it is the only chance offering, I respectfully recommend that the property be purchased on the most favorable terms possible, and that I be authorized to make the purchase.

A reply by telegraph will be very desirable as the storage space for goods is difficult to obtain at this place and the machinery from Richmond I am now expecting to arrive in a day or two. The train left Charlotte on last Monday.

The purchase of the proposed site for the Armory and the premises of Messrs. Pitts & Cook for the erection of the stock machines and for the purposes of a temporary machine shop will be in my opinion the very best that can be done; and I shall be glad to receive your instructions as soon as possible in relation thereto.

Should you agree with the purchase of the latter, I beg to request that you will detail Mr. J. Fuss, Master Builder at the Richmond Armory, and such of the late operatives of the stock machinery as may desire to come here—for duty at this point, and furnish them with free transportation together with their families. I shall also be glad to have the services of a Mr. Lotz, now employed in the Machine Shop at the Richmond Armory, and also those of some other machinist familiar with the erection of our stock machines who can be selected by the foreman Mr. Davis. These two last will also require transportation.

It would be well to require such workmen as may thus be sent here free of cost to themselves to sign an obligation to serve the C. S. Govt. at this post for a period of not less than 6 months. This course may have trouble hereafter.

Messrs Spiller & Burr have secured premises at this place which will answer their purpose and they now have workmen engaged in refitting the building.

Mr. McNeill also proposes to locate his carbine factory here also, and if so, this will become the great seat of small arms mfgy in the South." (Vol. 20)

## LOCATION OF THE PISTOL FACTORY IN ATLANTA

Burton comments upon Spiller & Burr having secured premises and that workmen were then engaged (June 11th) in refitting the building. This building was the old Peter's Flour Mill, located near the junction of Peters and Castleberry Sts. It is sometime referred to as being "where the Georgia Railroad Depot now stands," and sometime as just "near the Georgia Railroad Depot."

Wallace P. Reed's "History of Atlanta," states there was a steam flour mill of three stories, frame construction built in 1848 or 1852 by a Richmond Peters, W. G. Peters, J. F. Mims, and an L. P. Grant (the same "Grant" Burton saw on May 31st, the first day he was in Atlanta). When war broke out, the machinery from the flour mill was sold to the Confederate Govt., and was moved to Augusta to make gun powder. "The three story building was then used as a pistol factory." As there was only one pistol factory in Atlanta, this must have applied to Spiller & Burr. It was known as the "Atlanta Steam Mill." The Georgia Rwy. secured the parcel of land on which it stood on June 7, 1862.

While Spiller & Burr were busy in their new location, erecting and making machines etc. preparatory to the manufacture of their revolvers, Burton was still seeking the proper site for his Armory.

June 21st, he wrote Richmond that he had been disappointed with reference to the purchase of the old Race Track, and had abandoned the idea of locating at this particular spot. It appears the owners wanted $30,000 for it, a price considerably more than the shrewd Col. Burton figured it was worth. (Vol. 20)

By June 25th, he had decided that Atlanta was not for him, and that despite some disadvantages, Macon, Ga. was the more acceptable spot. He so advised his chief in Richmond.

## BURTON DECIDES UPON MACON AS HIS ARMORY SITE

Burton to Gorgas, June 25, 1862

"Since writing you last, I have again visited Macon at the request of several leading citizens of that city, and the result of my visit, together with certain circumstances which have occurred here, have greatly tended to influence me to reconsider my recommendations in favor of this place as a location for the new Armory.

When I left Atlanta last Sunday for Macon, I thought that I had made complete arrangements for the purchase of the piece of land mentioned in my last letter to you, and had employed a lawyer to look into the title, prepare papers, etc. On my return he informs me that a 'warranty deed' is refused to be given by the party who controls the property, and hence I have declined to consumate the purchase.

The fact of the business is, that the person who controls the property because of a mortgage on same, is a large property owner in this city, and it is my opinion that he desires to become the owner of the property eventually himself, and hence throws this obstacle in the way to prevent the sale to the Govt., and particularly as I have bargained for it at a fair rate.

This illustrates the prevailing feeling manifested by the people generally of this place towards the Govt. through its agents.

This of course forces me to look about again for means to set up the stock machines, and as events have transpired, I think they will prove to be for the best after all, as I will proceed to explain.

In the first place, the citizens of Macon are most anxious for the location of the Armory at that place, while quite the contrary seems to be the prevailing sentiment here. Upon the occasion of my late visit to Macon, I was called upon by one of the leading citizens who requested me to ride with him around the city, and examine the ground with reference to the requirements of the proposed Armory.

As I observed in a former letter, Macon offers superior advantages with respect to eligible sites and upon this occasion, I found one offering everything that could be desired for the purpose. The land belongs to the City of Macon, is well within the city limits, is sufficiently high to be healthy, and is both well watered, and adjoins a railroad. In fact it is a beautiful site for the purpose in view, and I doubt if another so favorable could be found.

But there is another and a great advantage offered in connection with it. The citizens, mayor and council very properly anticipating the advantages to be derived from the location of so large a Govt. establishment in it, have determined to do a most liberal act in order to secure these advantages.

I am therefore authorized to inform you that the City of Macon will make a free gift to the C.S. Govt. of about 30 acres of land, embracing the before-mentioned eligible site for the Armory, upon the condition that it shall be located thereon. When it is considered that this land is now worth $1,000 per acre, the liberality of the offer will be most manifest, and I at once took occasion to thank the City authorities in the name of the C.S. Govt. I thanked the gentlemen acting for the former for the very liberal proposition they offered, and which I stated I would seriously consider.

Now there is a good opportunity of accomplishing without cost what it seems I can not accomplish at this place at any cost. The objections to Atlanta, in this connection are increasing daily.

There are now several thousand sick soldiers in the hospital here, and the price of provisions have increased greatly within the few weeks I have been here. Flour is selling at from $20 to $23 per bbl., salt $50 per sack, etc. etc. I fear that mechanics cannot live here on such wages as ought to be paid them. There is difficulty also in obtaining houses owing to the unusual influx of strangers from the coast and elsewheres. There are little or no facilities here for building or doing work of any kind, and no materials as I before mentioned.

In all the above particulars, Macon has the advantage, and in view of facts as they now exist, I without hesitation am induced to recommend that the liberal offer of the City of Macon be accepted, and that steps be at once taken to commence operations there. I have done nothing here to commit me in the least to this place, and the work can go on in Macon as well as it could here. The bill of timber that I have ordered will only have to be transported there from here when it is delivered.

Fortunately, I have kept all the stock machines in the cars here so that they can be at once sent on to Macon without trouble, and I shall be able to erect them quite as soon in Macon as in this place.

I therefore respectfully recommend that the offer of the City of Macon be accepted by the Govt., and that you instruct me to locate both the Armory and the stock machines at that place. I might state other advantages to be derived from the choice of the latter place, but I deem it unnecessary as those I have stated will be sufficient. I do not doubt to induce the Govt. to decide in its favor. I confess that I am disappointed in Atlanta as a location for the Armory and my last letter will have prepared you for some recommendations similar to that I now make. Speculation in real estate seems to be the sole object in view by the citizens of this place, and hence the indifference manifested towards the establishing of manufacturing and other enterprises of industry. This state of things must change before Atlanta can become a thriving city." (Vol. 20)

The property offered the C.S. Government by the City of Macon, consisted of a lot bounded by Calhoun, Hazel, Lamar, and the Western Rwy. tracks, and an additional lot bounded by Calhoun, Ash, Elm and Ross Sts. This land was titled and transferred to the C.S. Govt. by act of Georgia Legislature in 1862. (page 44 Acts of Georgia Legislature)

Burton was given the "go-ahead" by Gorgas, and there-after devoted his seemingly boundless energies towards the erection and establishment of his "Armory."

## OPERATIONS AT THE C.S. ARMORY, MACON

His activities in this connection are covered in more or less detail in the rear of this book for those interested, but for a quick look at the Armory, and its operations as of Jan. 7, 1863 we turn to a letter from Burton to Major Gen. Ben. Huger, Inspt. General of Ordnance & Artillery, C.S.A.

"I have the honor to submit for your information, the following statement in relation to the operations of the Armory at the present time.

*Officers on duty*
Jas. H. Burton—Supt.
Charles Selden, 1st Lieut.—Paymaster & Masterstore Keeper
*Subordinate Officers*
Jeremiah Fuss—Act. Master Armorer
Augustus Schwaab—Architect & Engineer
William Copeland—Master Machinist
*Clerks*
John Allen—Clerk to Paymaster & MSK
Henry C. Day—Clerk to Supt.
Daniel E. Stipes—Clerk to Master Armorer
*Foremen*
George W. Weston—Foreman of Stock Dept.
Wm. H. Thornberry—Foreman of Mach. Shop
Oliver Porter—Foreman of Laborers
B. F. Perry—Foreman of Carpenters

*Operatives*
Mechanics—white 45
Mech. apprentices—white 7
Watchmen & laborers—white 5
Mechanics carpenters—negro 23
Laborers—negro 56
    total—136

## WORK DONE

The work at the present time in this armory consists in the preparation of gun stocks by machinery. The fabrication of machinery for the manufacture of arms—and other such work of a general nature required in connection with the erection of permanent buildings for Armory purposes.

The number of gun stocks prepared and forwarded to Richmond, Va. is about 1,500 per month, to produce which, the machinery is run every night until 8 PM, and occasionally on Sundays.

*Wages*

| | |
|---|---|
| Average wages of machinists | $3.25 per day |
| Average wages of carpenters—white | 2.50 per day |
| Average wages of laborers—white | 1.50 per day |
| Average wages of carpenters—negro | 1.85 per day |
| Average wages of laborers—negro | 1.25 per day |

## REMARKS

Much difficulty is experienced in consequence of the very limited number of good machinists and blacksmiths available. The shops and machine shop machinery and tools are of sufficient capacity to employ 170 machinists and blacksmiths. At the present there are but 34 machinists, and 2 blacksmiths employed. Every effort has been made by advertising and by other means to increase the number, but with little success. The demand for the class of labour so greatly exceeds the supply that it is in vain to expect to meet it, and to a great extent, this class of mechanics dictate the wages they receive because of the limited competition among them. Applications in the form of petitions are made almost monthly by those employed in this Armory for increase of wages. This difficulty points to the necessity of providing some other means by which to obtain the large amount of machinery necessarily required for so large an armory. I respectfully suggest and recommend that an effort be made at once to have it fabricated in England." (Vol. 31)

The "stock machines" to which Burton frequently refers were captured by the State of Virginia in April 1861 at Harper's Ferry Arsenal. They were first brought to Richmond, Va. as Virginia State property. Later they were "loaned" to the Confederacy for the duration of the war. On June 12th, 1862, this machinery was sent on to Atlanta, Ga. in a 5 carload lot, and were finally forwarded on to Macon. It was these machines which provided the stocks for the arms manufactured at the Richmond Armory, and for the Richmond "Sharps." Thus it can be seen that their importance can not be over estimated.

## THE C.S. ARSENAL AT MACON

Burton's Armory, was not the only government establishment at Macon. Also located at this point was the C.S. Arsenal & Laboratory, which was also a large operation. Says the "Historical Record of Macon" (page #258);

Right and left side of The "Sample" Spiller & Burr Revolver *(from the collection of the author)*

"In May 1862, the Arsenal at Savannah was removed to Macon, Ga., under the direction of Col. R. M. Cuyler and the large foundry of the Messrs Findlay (J. D. & C. N.) was appropriated to ordnance. From 350 to 500 hands were employed in the manufacture of cannon, shot, shell and harness. The 12 pounder Napoleons made here were the pride of the army."

## C.S. ARSENAL ATLANTA

Despite the future potential of the Armory, Arsenal and Laboratory at Macon, the largest and most important Confederate establishment south of Richmond, was the C.S. Arsenal in Atlanta, Capt. M. H. Wright Comdg.

Capt. Wright, a former West Pointer, was an exceptionally capable officer. He entered the Confederate service as a lieutenant in 1861, and by 1862 was promoted to captain, and before the year was out was made a major.

The records of the Atlanta Arsenal are reasonably complete, and show without question that no arms were actually fabricated there, although considerable alteration and repair work was done, and the Arsenal had its own armory and laboratory.

To this establishment were funneled most of the arms and ordnance supplies from Macon, Ga., Columbus, Miss., Columbus, Ga., Nashville, Tenn., Dalton, Ga., Selma, Ala. and Augusta, Ga. The Arsenal acted as an arms and ammunition depot for the Army of Tenn., holding the same position for this army that Richmond did for the Army of Northern Virginia. Stores, ammunition, accouterments, arms, etc. were issued as required.

A letter to the Provost Marshall, Atlanta, Ga. on March 15, 1863, lists the following buildings as being occupied by the C.S. Arsenal: Bldg. #1—Military Storehouse, corner Peachtree and Walton, Bldg. #2—Armory & machineshop, corner Decatur and Pearce, Bldg. #3—Finishing Dept., on Pearce St., Bldg. #4—Laboratory Pyrotechnical Dept., on Washington St., Bldg. #5—Cap-forming Dept., on Mitchell St. 2 down from Whitehall, Bldg. #6—Harness & Saddle shop, Whitehall over Henrietta, Marchley to Joiner, Bldgs. #7 to #13—Laboratory together with magazines on old race course. (Vol. #10)

The Atlanta Arsenal then was a sizable operation. Remembering this, and also its importance to one of the major armies of the Confederacy, it is of interest to see what this arsenal had "on hand" in the way of weapons. Such a look, might give us an insight into Confederate Ordnance.

A statement from the Arsenal dated Feb. 21, 1863, shows the following arms were to be found there: 1,116 flint lock muskets, 1,132 percussion muskets, 255 rifles, 11 cadet muskets, 117 assorted carbines, 355 double-barreled shot guns, 38 single barreled shot guns, 208 rifles—cut off & bored out, 214 unfinished rifle barrels, 217 Enfield barrels (rifle), 68 rifle musket barrels, 132 musket barrels, 318 sword bayonets, and 1,111 bayonets. (Vol. #10) One would be safe in saying that this was not a large assortment with which to fight a war, and a complete absence of revolvers and pistols is noted.

Oladowski, Chief of Ordnance for the Army of Tenn., suffered for the lack of these revolvers, and so informs Wright, who is forced to reply March 3, 1863: "I have not a pistol for issue to the cavalry, but will send up a few carbines when they get the road open." (Vol. #10)

## SPILLER & BURR IN ATLANTA

Where were all the revolvers, and particularly, where were the 15,000 pistols called for by the Spiller & Burr contract?

Unfortunately for Wright, for Oladowski, and for the hardpressed Confederate Army, the Spiller & Burr plant's product was still in the stage of "forthcoming."

From the time of the firm's removal from Richmond to Atlanta in May, 1862 until the fall of the same year, available records are silent as to the exact extent of their activities, and during this period only fragmentary reports of them come to light:
July 15, 1862, Lt. Col. Burton, C.S. Armory, Macon to E. N. Spiller Esq, Atlanta.

"Sir, I will thank you to loan me for the purposes of this Armory, your set of patterns for cast iron forge, and all that relates to it. Please have them packed, and forwarded to my address here. The Govt. will pay you for the use of them." (Vol. #20)

Sept. 29, 1862: "Wanted a brass moulder. A 1st class brass moulder and finisher is wanted to go South. Address or call on David J. Burr, at the Life Insurance Office, corner 10th & Main Sts., Richmond, Va." (Richmond "Dispatch.") This ad would indicate that Burr remained behind and did what he could to assist operations from the Richmond end.

Oct. 21, 1862: "Wanted to rent in Atlanta, a comfortable dwelling. Rent to commence at any time between this and the 1st of Jan. 1863. Address thru Post Officer, or call on E. N. Spiller at the Pistol Factory, Atlanta." (Richmond "Examiner")

Oct. 22, 1862: "W. N. Bingham, on trial for taking a negress from her owner in Macon, Ga. He was released when he showed he was an employee of Spiller & Burr, contractors for the manufacture of small arms, then moving to Atlanta, and had applied for transportation." (Richmond "Examiner")

From November 1862 on, we are able to follow the firm's activities somewhat closer.

Nov. 29th, Spiller writes Burton: "I wrote you some time ago in which I made a request that you come up this week. As you have failed to do so, I write again as I think it necessary you should see the rifling machine on which all the ingenuity I have brought to bear has been used without accomplishing what it should. As it is, I do not think it will do, although it may be changed so as to work. There is too much twist in the grooves for one thing I think. I am sure that 5 instead of 7 rifles would do better. This machine is the only one in which we are likely to have any delay."

The above, written one day before 4,000 revolvers were to be delivered to authorities in Richmond (according to the terms of their contract), reflects very plainly that revolver #1, (let alone 4,000) was still unfinished.

## SAMPLE REVOLVER

By Dec. 17th however a letter from Col. Burton to Col. Gorgas, Chief of Ordnance, indicates the firm were at last finishing their "sample-revolvers" which they would exhibit shortly there-after.

"Mr. E. N. Spiller of the firm of Spiller & Burr will arrive in Richmond about the time you receive this, with a sample of their manufacture of pistols, and with which I trust you will be pleased. Whilst the sample is in my opinion the best that has yet been produced in the Confederacy, yet it is not quite up to my standard of excellence in several minor particulars. These will be improved as manufacture progresses until the desired degree of excellence is arrived at. The contract of Spiller & Burr provides for the inspection of their pistols by a system of tests which have not yet been elaborated. I think it not only proper, but necessary that this should be reduced to writing without delay as required by their contract; and if you will authorize me, I will draw up a scheme and submit it to you for your approval in order that both parties to the contract may know beyond the possibility of dispute what rules shall govern the inspection.

The factory having been removed from Richmond at which place their contract provides for the delivery and inspection of their arms, it will be both inconvenient and unnecessary in my opinion to send the pistols all that distance for inspection. If I could think that you would not regard the suggestion as an improper one, I would recommend that the inspection be placed under my direction and that the arms be sent to me for that purpose at the expense of the Govt. I am aware that this is a delicate suggestion but my desire is to secure to the Govt. a *first class* weapon and I feel this can best be accomplished by my controlling the inspection. If the result is not satisfactory to the Govt., I am willing to assume the responsibility and this I would not do unless I felt sure that I could do the Govt. justice. I but do myself justice when I say that I have had some little trouble in the endeavor to keep Mr. Spiller on the right track in connection with this contract. The misfortune is that he has been bred to purely commercial pursuits and consequently he is to a certain extent 'at sea' in manufacturing operations. He is rather too much inclined to speculate and make the most of his chances and it is this spirit I have endeavored to restrain almost at the expense of friendship between us. But I have restrained him so far and if you will sustain me, I will continue to do what I think is right for both Govt. and contractors.

The letter I wrote you some weeks ago on the subject of contractors undertaking work foreign to their contracts with the Govt., originated in the desire of Mr. Spiller to undertake to construct machinery for other parties when I knew that his own work would suffer in consequence. I opposed him in this desire, and he did not undertake the work. He is now desirous of making his pistol barrels of iron instead of steel with a view of disposing of his stock of steel at the present high prices. He has steel enough on hand, bought mainly one year ago for all the pistols he is under contract for, and my desire is that you will insist on the barrels being made of steel. The cylinders must be of iron as steel can not be obtained; but by twisting the iron, the fibres can be thrown in a direction around the circumference of the cylinder and the requisite strength thus secured. Mr. Spiller has energy and perseverence, but requires the influence of some competent governing power to keep him in the right groove in connection with his contract with the Govt., and I mention these facts not for the purpose of injuring him in your estimation but to show you that it is my earnest desire to honestly fulfill a contract in which I have a considerable private interest.

The pistols submitted by Mr. Spiller have been made after a pistol obtained from the U.S., which has so far served as a model, but I think it will be well for Spiller & Burr to now get up a model pistol to serve as a standard of reference, the cost of which the Govt. should defray, as also that of a set of inspection guages. All this I will attend to, if you will authorize me.

I have some doubt as to the calibre of these pistols being the same as that of the Colt's Navy Pistol which I think may be a trifle larger. As I have no means of testing this exactly, I mention it in order that you may direct the point to be determined. I think the calibres should be the same, and now is the best time to make the correction if required. I shall be glad to hear from you when you have the leisure to write."

## TWISTED IRON FOR CYLINDERS

One of the most interesting points of the foregoing letter is contained in the statement: "The cylinders must be of *iron* as *steel* can not be obtained, but by twisting the iron, the fibres can be thrown in a direction around the circumference of the cylinder and the requisite strength thus secured." A close visual examination of the Spiller & Burr revolvers will show the twisted iron cylinder, a peculiarity which extends to other Confederate revolvers (Griswold & Gunnison) and a point which collectors might bear in mind where faking is suspected.

Few persons can realize just how desperate was the South in her need for metals. Possibly the following quote from Caroline Jenkin's diary, dated March 3, 1864, will show to some extent what this lack meant even in the life of the everyday citizen:

................"....One hears with a sinking heart of the dreadful shortage of metal, so much needed for bullets. Another call has come through for lead and steel. With the house already stripped I can think of nothing to send except kitchenware, of which there is little enough in all conscience. I will send what knives I have that have steel blades and a coffee grinder which I shall not miss as there is no coffee to grind."

### DESCRIPTION OF SAMPLE REVOLVERS (See photo)

Jan. 1, 1863, Spiller returned to Atlanta, after a visit to Richmond, Va., where he had shown his "samples" to the War Dept. Please note as regards these samples that reference is consistently made to them in the plural, so several must have been made up.

In my possession is a brass-framed Whitney conforming to those made by Spiller & Burr. It is an exceptionally well made weapon showing care and precision in all respects. The grips are of burled walnut, carefully finished, the frame is well cast, showing no "faults." Instead of the customary brass pin, the foresight is a steel pin carefully fashioned and mounted on a small metal plate which is inlet into the barrel. The cylinder is of twisted iron, the lateral "faults" showing clearly. This revolver is totally without markings. It contains no name, no serial number, and no inspector's stamp, but comparison with many Spiller & Burr revolvers, leaves little doubt but that it is a product of this pistol factory. The conclusion reached is that this might be one of their so-called "Samples."

### SPILLER SUGGESTS A NEW CONTRACT

Spiller to Burton, Jan. 1, 1863.

"I reached home from Richmond on Tuesday night, having had a very disagreeable trip both going and coming. Faired very well in Richmond. Our pistols were very favorably received. Col. Gorgas seemed very well pleased with them indeed. Maj. Downer inspected them, and gave a very flattering report so Butler informed me, he having seen the report before the major sent it in. He said in many respects they were the best arms that had been turned out in the South. He also recommended doing away with the silver plating. The Col. suggests making the catch spring of the loading lever like Colt's, which I told him we intended to propose doing.

We shall have to make them Colt's calibre, which I find is a small fraction longer than ours: it will give us some little trouble as we shall have to make some changes. In regard to advance in prices; the subject was opened and favorably viewed by Col. Gorgas. I referred him to the immense difference in prices of everything entering into our work and at his request made a written statement in regard to the matter with some facts connected with our enterprise which he said he would place before the Sec. of War with his own views in regard to the matter. I think from indications we shall at least get the advance we spoke of, if not more. He said ours would be the first contract acted upon and would be the basis of the advance on other contracts. He suggested that such an advance should be made as would include the advance agreed upon on account of our removal as it would make the matter less complicated, which I told him would be perfectly satisfactory provided the whole advance was sufficiently liberal. I am anxious to hear in regard to the matter, and will advise you as soon as I do. Col. Gorgas desired to be remembered to you.

Butler has not returned yet. He came by Fayettesville to see his mother, and I hope he will get here this week. He was quite sick on his way to Richmond from a cold. Helen is doing finely and does not care to come home. All are well and desire to be remembered to you and family. Let me hear from you."

The "Butler" that Spiller refers to so often in his correspondence was Reese H. Butler. Butler was formerly Foreman of the Bayonet & Mountings Dept., of the Richmond Armory, but left this job in order to follow Spiller & Burr South. His title with the firm was that of General Manager, and he had the complete confidence of his employers.

In the letter we have just quoted, Spiller states that his "sample pistols" were favorably received and inspected. He had every reason to be pleased with the inspection report prepared by Major Downer, Comdg. Richmond Armory, as witness:

### INSPECTION REPORT ON THE SAMPLE SPILLER & BURR REVOLVERS

"Ordnance Depot
Richmond, Va.
Dec. 26, 1862

Col. J. Gorgas, Chf. of Ordnance
Richmond, Va.
Colonel:

I would respectfully report that I have examined the pistol made by Messrs. Spiller & Burr & Co., critically and find no defects in them which will not remedy themselves as the machines and tools become adapted to the work required, except as are incidental to the model. I think the style of the catch to hold up the lever of the ramrod is faulty as the spring is necessarily weak, and after wearing awhile, it will become worthless for the purpose for which it is intended. I would recommend a spring and catch like that of 'Colt's' pistol. I would also recommend a slot cut in the base of the cylinder between the cones in which that face of the hammer will fit, holding the cylinder at a half revolution and making a safeguard from accidental explosion. I would also recommend the adoption of the calibre of Colt's Navy Revolver for the sake of uniformity in ammunition. The calibre of this pistol is somewhat smaller than Colt's. The present cylinder and barrel will bear the necessary increase. I think rounding of the muzzle of the pistol is an improvement as well in appearance as in use, as it is less apt to cut the holster or the sharp edges to become bruised.

I find the workmanship on the pistol to be of a character in the highest degree creditable to the makers—much of it exceeding in quality that of the model. I would beg to suggest however that I think a plain brass mounting is superior to plated—as the plating soon rubs off on the parts most handled—

giving the work a mottled and unsightly appearance.

Respectfully etc.<br>
W. S. Downer<br>
Supt. Armories" (vol. #90)

## INSPECTION REPORT OF THE LEECH & RIGDON REVOLVERS

Major Downer also inspected the revolvers made by Leech & Rigdon. This report is not quite so favorable;

"I respectfully report in regard to the pistol presented by Messrs. Leech & Rigdon, that I find it a good and serviceable weapon, and worthy of the patronage of the Department.

The faults I notice are: want of uniformity in the calibre in the chamber, and an inferiority of the iron in the barrel. I also note that the parts do not interchange which they should do. In regard to the proposal for a contract which they offer, I would suggest that their terms are only liberal to themselves, but aside from that, the state of the country is such in regard supplies, labor, etc., that I think we would be impolitic to enter into a contract of such size now. I would recommend, however, that all the pistols which can be made by this firm within a year from this date, or say during the war, be purchased by the Dept. at the price named, and a small advance made on approved security, they entering into contract to deliver to the Dept. all the products of their factory." (vol. #90)

This report is dated Jan. 21, 1863, at which time the firm of Leech & Rigdon were still at the Briarfield Armory, Columbus, Miss., and is conclusive proof that they were manufacturing at that point, and that their first revolver was not turned out at Greensboro, Ga., as has previously been supposed.

## PROPOSED SYSTEM OF INSPECTION & TESTS FOR SPILLER & BURR

Upon receipt of Major Downer's report on the Spiller & Burr revolvers, Burton wrote to Col. Gorgas on Jan. 22nd;

"In compliance with your instructions of the 29th ult., I have prepared the enclosed paper defining the system of inspection and the tests to be applied to the pistols to be supplied by Messrs. Spiller & Burr of Atlanta, Ga., on their contract with the War Dept., and according to one of the provisions therein stated, you will observe that the system of inspection is a rigid one, and such as would be quite sufficient to ensure good arms in times of peace; but at the present time may have to be relaxed somewhat. I shall endeavour, however, to secure work of a quality as nearly approaching the standard as it is possible to attain. Should you approve of the system as I have drawn it up, I respectfully suggest that you cause a copy to be prepared and signed by yourself and either forward it to Messrs. Spiller & Burr direct, or through me—as you may deem most expedient.

Under the heading of the 'Barrel,' you will observe that the established diameter of the bore of the barrel and the maximum limit allowed *layer*—are left blank. As I have no pistol here by which to arrive at these dimensions satisfactorily, I respectfully request that you will cause the measurement to be made, and reduced to figures at the Richmond Armory, and the blanks filled up so that the limit shall be .002 of an inch larger than the standard diameter of the bore, and inform me of the same.

A bullet mould and screwdriver should be supplied with each pistol.

Shall I prepare moulds for them, and agree on a fair price for the supply of these implements, as provided for in the contract?

P.S.—The better plan will be to prepare and send me by mail a small gauge plug for the standard size of the bore of the barrel as you establish it." (vol. #20)

The following day (Jan. 23rd) Burton writes to Spiller & Burr;

"Gentlemen:

I am instructed by Col. Gorgas, Chief of Ordnance, to request you to prepare as soon as possible, a model pistol of the pattern you have commenced to manufacture, and which shall be fitted with a catch for loading lever in all respects like that attached to 'Colt's revolvers' navy size, and the calibre of the barrel of which shall also be of the size of the latter named arm, a standard guage plug for which I have requested to be forwarded to me from Richmond, and a duplicate of which I will send you as soon as possible. This model is required by the Govt. as a standard of reference and also for the purpose of preparing guages for the inspection of the arms supplied by you and therefore it should be as perfect in all respects as it can be made. The cost of fabrication, provided it be not excessive, will be defrayed by the Govt., as it is intended to become the property of the Govt. It would be well for you to prepare a duplicate at the same time for your own use.

As soon as you have any pistols ready to deliver, you will please forward them to this Armory (Macon) for inspection, and I trust that the first lot will come forward soon. Any number less than one hundred will be inspected if offered, although it will be preferable to have them delivered in lots of not less than 100, as provided in your contract." (vol. #20)

Evidently Gorgas felt that Burton, in Macon, was too far away for the direct supervision necessary for Spiller & Burr in Atlanta, and to overcome this difficulty, Wright of the Atlanta Arsenal was given supervisory authority over the pistol factory. Feb. 3rd, Wright therefore had the occasion to advise the contractors:

............."Gents, I have the honor to inform you that the supervision of your contract with the C. S. Govt., has been confided in the comdg. officer of this Arsenal. In making application for the detail of men to carry on your work, you will apply to the Comdg. Officer of this Arsenal, accompanying the application with the affidavit as required in P. 1X, Gen. Order #22, A & I General Office, an extract of which is enclosed." (vol. #10)

*A NEW CONTRACT FOR SPILLER & BURR, BUT STILL NO PISTOLS*

The two letters which follow indicate; #1—the new contract between Spiller & Burr and the Confed. Govt., #2—some of their difficulties in production, and #3—that as of Feb. 14, 1863, serial #1, had not left the factory. E. N. Spiller to Col. Burton Jan. 27, 1863:

"I have at last received from Col. Gorgas the new form of our contract with an advance on the price of the pistols. It does not meet my views exactly, but I do not know whether we can get it bettered or not. The raising of the Blockade should not have anything to do with our contract it strikes me: the part of the work that bears hardest on us has mostly been done—indeed, I can not see that we should be benefited. $25,000 in every way from this to the completion of our contract—should the Blockade be removed in a moment and yet according to the provisions of the contract, we should be damaged over $150,000 if the Blockade is raised in a month or two, which may be the case. I can not see that our prospects are much better by this than the other contract. In the first clause of this contract he makes an error in saying we contracted to make 'Colt's Navy Pistol.' The electroplating is left out I am glad to say, although I had made, or almost completed the most perfect arrangements for that process and they would have cost .50 cts. for each lock frame. Look over this thing and give me your views as early as possible. I am in such a condition as hardly to be fit for anything. Fear I am much worse than when I wrote you and don't know what is the matter. I can not sit up much.

I am enclosing a copy of the contract sent me by Capt. Smith of the Ord. Office."

E. N. Spiller to Burton, Feb. 14, 1863:

"Yours of the 13th inst. is just to hand.

In regard to the wire, I think I will not buy it, as Butler bought a lot in Fayetteville when there some weeks ago.

We have no pistols out yet, but I think will get some hundred together by the end of Feby. It shall be as soon as possible. I am worried out of my life by the delay. I would not go through the worry and bother that I have so far again for $20,000. Butler does his best, also some few of the hands, but most are utterly sorry and unreliable. We are not doing as well as we can with what we have to do.

I have a great many bothers to contend with and the most infernally difficult place to get anything done in this world. I will come down to see you as soon as I can get some pistols ready. After being inspected by you, who is to receive the pistols? Maj. Wright I suppose, as he is appointed to supervise our work here.

I enclose you the draft of what I have determined to adopt as to the loading lever. It is an improvement I have lately seen on a new pistol taken from the Yankees, and the most complete I have seen. Works as well or better than 'Colt's,' is handsomer and can be much more easily made than 'Colt's.' The catch for the end of the Loading Lever is clamp milled round, having a neck for a screw which is screwed into the barrel, having first had one side milled flat for the hole to receive the point of the 'L' lever. The connection you will see between the lever and rammer is simplified, being made like 'Colt's,' and obviating the link. I should have written you sooner in regard to this, but that I thought until a day or two ago, I should come to see you. Let me hear from you sooner."

## SPILLER MEDDLES WITH OTHER PISTOL CONTRACTORS

Despite his own difficulties, it seems that Spiller was not too busy to concern himself with the operations of other pistol makers. On Feb. 2, Col. Gorgas saw fit to write Burton; "—Spiller requires a word of caution too; to keep strictly to the contract and to avoid meddling with the hands of other establishments of arms."

This reference was in connection with the Haiman Revolver Plant at Columbus, Ga. Burton had already called Spiller sternly to task for furnishing this concern with plans and specifications which Burton considered his own personal property, and only "loaned" to Spiller & Burr. However, as L. Haiman & Bro., proprietors of the Columbus Firearms Co., turned out imitation Colts, not Whitneys, the plans and specications probably referred to a milling machine, rather than the guns themselves.

## MARCH 1863, AND STILL NO REVOLVERS

March 5, 1863, and the long awaited revolvers of Spiller & Burr were still not forthcoming. Burton complains bitterly as to their delay.
Col. Burton to E. N. Spiller, March 5th:

"I infer from the purport of a letter I have received from Col. Gorgas dated 24th Feb., that up to that time, the proposed new contract with Spiller & Burr had not been duly executed. Please inform me if you know why the delay has occurred as I am at a loss to account for it in view of the length of time that has elapsed since you received the contracts to sign. Please bear in mind that Col. Gorgas holds *me* responsible for the proper working out of this contract as stipulated and excuse me therefore if I seem overanxious on the subject.

As I have said before, I think you ought now to be in a condition to produce arms in quantity and as they are not forthcoming, I shall feel constrained in justice to myself to express an opinion to this effect to Col. Gorgas in order that I may be relieved of the responsibility of further delay.

I do not wish to do this, but as I am expected to, and do know what ought to be accomplished in such business, Col. Gorgas looks to me to explain the cause of delay, which with the light now before me, I candidly confess I cannot do satisfactorily. Col. Gorgas hints also at the interference by you with Govt. employees at other establishments. I do not know if he has just grounds for hinting at such a thing, but you must be aware that such interference will not be tolerated if attempted. I deem it proper to inform you on the subject in order that you may guard against it in the future.

I shall be glad if you will specifically explain the difficulties in the way of your progress at the present time in order that in addressing Col. Gorgas on the subject I may state them for his information."

## THE NEW CONTRACT

Although Burton's letter is dated March 5th, and questioned as to why the proposed new contract had not been executed, as a matter of fact, this contract had already been signed as of March 3rd., it superseding the old contract which was abrogated by mutual consent.

The new contract called for 600 pistols to be delivered in Feb. 1863, and 1,000 a month thereafter until 15,000 had been delivered. Payment was to be $43 for the first 5,000, $37 for the second, and $35 for the final 5,000. These advanced prices included reimbursement for the removal of the factory from Richmond to Atlanta.

The terms of this new contract are quite explicit, but one point needs clarification—how in March 1863 (when no revolvers had yet been produced) the firm was to make delivery of 600 revolvers in Feb. of the same year? We can only surmise that the agreement was drawn up prior to Feb. at which time a number of revolvers were near the finished stage.

After posting the previously quoted letter, Burton must have received word regarding the new contract, for on the same date (March 5th) he writes his second letter to Spiller;

........................"Col. Gorgas, Chief of Ordnance, has called my attention by letter, to the fact that by the terms of the new contract he has made with you, 600 pistols were to have been delivered during the month of Feb. ulti., and I am charged with the inspection of them according to instructions, a copy of which will be furnished to you when you make your first delivery.

As no pistols have yet been presented by you for inspection, it is my duty to call your attention to the subject with a view of ascertaining from you when you will make a delivery, and in what number?" (vol. #20)

### STANDARD GAUGE FOR THE REVOLVERS

Burton's next letter to Spiller & Burr (March 14th), is somewhat milder in tone.

"Enclosed, herewith please find system and inspection for your pistols which has been approved by Col. J. Gorgas, and to which your assent is requested. You will receive also by mail, a parcel containing a compound gauge plug of 3 diameters which is to establish the dimensions (calibre) of the bore of the barrel, and that of the cylinder. Please preserve this gauge, as a standard, and make working guages from it. The plug marked '.3675' is for the bore of the barrel at the muzzle; that marked '.3680' is for the bore of the barrel at the breech above the coning—that marked '.37' is for the chambers in cylinders.

I find the bore of the barrel of the pistol you sent to me very close to gauge at the muzzle. The chambers of the cylinder are rather small to gauge and will have to be enlarged slightly. As soon as you furnish me with the model pistol, which I beg you will do as soon as possible, I will prepare the necessar gauges for the inspection of your pistols." (vol. #20)

### NEED OF SKILLED WORKMEN NECESSARY

Col. Oladowski, Chief of Ordnance, Bragg's Army, was still in urgent need of revolvers, which fact he communicated to Col. Gorgas in Richmond, who in turn passed on the communication to Burton. Here was a chance for that ingenious colonel to kill two birds with one stone—to get skilled workers in the pistol factory, and also to start turning out the long delayed revolvers. With this thought in mind Burton addressed Col. Oladowski (April 18, 1863);

"There is now in Atlanta, Ga. a factory with complete machinery (for which I furnished drawings) capable of producing 1,000 pistols per month, but in consequence of the want of sufficient and competent workmen, but a small portion of that number is turned out. I write at the suggestion of Col. Gorgas to say that if you can manage to detail 10-20 good machinists who are competent to do the finishing work on pistols, and order them to report to Messrs. Spiller & Burr at Atlanta, you will soon be supplied with pistols of superior description. I am charged with inspection of all pistols made by these parties, who have a large contract with the Govt., and I am daily in expectation of a small lot, say 100 which doubtlessly will be sent to you.

I shall be glad to know from you that you can detail workmen to the above named contractor, as by this means, your wants in this particular can be best soon supplied." (vol #20)

As the Atlanta Arsenal records reflect that men were detailed to the Spiller & Burr plant from Bragg's Army, it is reasonable to suppose that the bulk of the revolver output, when it finally appeared, was issued directly to Oladowski for cavalry usage. To substantiate this line of thought is a communication between Wright to Gorgas (March 13, 1863);

............................................."I have the honor to report that Messrs. Spiller & Burr will have shortly, a large number of pistols ready for inspection. I understand that Col. Burton at Macon is ordered to inspect the arms. Will you have the kindness to instruct me whether arms are to be sent to Richmond, or to Gen. Bragg's Army, where arms are badly wanted for cavalry?" (vol. #10)

While Bragg's cavalry was undoubtedly delighted to receive anything in the way of a firearm, a later correspondence between Wright and Gorgas indicates that the troopers in the Army of Tenn., did not regard the brass-framed Whitneys in too high esteem;

.............................."I have the honor to transmit herewith duplicate requisitions for 200 pistols to supply the Army of Tenn. Col. Oladowski requests some imported pistols, if possible, Kerr's pattern in preference to those of the Spiller & Burr model, but if no imported ones can be obtained, he would be satisfied to obtain the latter." (vol. #16)

.................................................................Or in other words, anything was preferred over the nothing he then had.

## REVOLVERS FINALLY FORTHCOMING!

It was May 1863 before the first lot of revolvers left the Pistol Factory in Atlanta to be inspected by Burton in Macon. This lot consisted of a group of 40. The inspection was distinctly disappointing. Says Burton (May 2nd);

"The 40 revolving pistols manufactured and presented for inspection by you have been inspected and I enclose herewith a copy of the inspection report made by the inspector. You will observe on reading it that but 7 pistols have been accepted out of the 40 presented, and those 7 are by no means as perfect as they should be. The inspection of the others was not proceeded with for the reason that they presented 2 serious and fatal defects, viz; the chambers are not in line with the barrels, and there is much too great an allowance of space between the ends of the barrels and the face of the cylinders through which gas can escape. These defects could not be corrected here, and I have therefore given directions to the MSK to return the 33 pistols to you, and they have been forwarded today by express.

I greatly regret that the inspection of these pistols has been so unsatisfactory both on your account and in view of the wants of the Army in Tenn., in which pistols for cavalry are much needed. The inspection of these pistols however defines the defects, and will I trust, enable you to adopt means to overcome them in the future. The 7 pistols accepted will be retained here until others are received, as the number is so small that it is not worthwhile to forward them to Major Wright unless you specially desire it." (vol. #33)

Mind you now, this is May 1863! Thus far, the total output has been 7 revolvers. Remember the contract of Nov. 1861 called for delivery of 15,000. Oh well, only 14,993 to go!

May 7, 1863, Col. Burton left Macon for an extended trip to England. The newspaper article quoted in his obituary indicated this trip was made for the Dept. of State. Actually, he went under direct orders from Col. Gorgas for the sole purpose of buying ordnance, machinery, tools and equipment.

During Col. Burton's absence, Gorgas assigned Richard M. Cuyler to the command of the C. S. Armory, Macon, Ga. (Order #11, vol. #49), and during the several months that Burton was abroad, Cuyler was in command of both the C. S. Armory, and the C. S. Arsenal in Macon.

A few days after assuming command of the Armory (which had jurisdiction of the Pistol Factory in Atlanta) Cuyler wrote Spiller & Burr;

............................................................... "I notice a memorandum left by Col. Burton that you are required to make a model or standard pistol from which a set of inspection or standard gauges are to be made at this Armory. I desire to know if any steps have been taken towards making this model?" (vol. #31)

## SUMMER OF 1863

After the initial difficulties incidental to manufacturing had been overcome, the Spiller & Burr revolvers were forthcoming with some degree of regularity, but only at a trickle of 50 to 75 per month, instead of the 1,000 per month promised.

Apparently overwhelmed or unable to cope with wartime manufacture, dissatisfied with the new contract, and with their "spark-plug," Col. Burton in Europe, the firm of Spiller & Burr prepared to throw in the sponge, and suggested that it might be more profitable for all concerned if the Confederate States of America would step in and buy their plant, lock, stock and barrel.

This suggestion was verbally put to Capt. M. H. Wright, of the Atlanta Arsenal in early June 1863, following which on June 16th, he requested of the firm, certain information;

............................"Gentlemen, you will please make me a written proposition of the terms upon which you will dispose of your stock and machinery, comprising the Pistol Factory, stating: 1; kind, number and class of lathes and other machinery, 2; capacity of shops and machinery, 3; material and conditions of it, 4; progress of work, 5; terms and prices." (Vol. #11)

Evidently no agreement could be reached along the line of direct sale, and Spiller & Burr next proposed a revision of their second contract as a means of relief. This proposition they placed before Wright, who forwarded same to Gorgas on July 16th with this covering letter;

............................................."I have the honor to enclose herewith, the proposals of Messrs. Spiller & Burr, Govt. contractors, for a revision of their previous contract for revolving pistols.

I am not aware what pistols are coming into the Govt. in other districts, nor do I know what progress is being made in this manufacture, but I have no doubt that as the contractor proposes, the pistols will come cheaper than if they are brought from abroad, and believing the pistol contractors to be honorable and faithful men who have done all they could, and that the price is fixed by them on a basis of reasonable profit. I endorse this proposition accordingly." (vol. #11)

Apparently, Capt. Wright was favorably disposed towards the problems, etc. encountered in pistol manufacturing, but nevertheless in Aug. finds occasion to write the firm rebuffing them for seeking to induce workmen from Govt. establishments by offers of higher wages, and warns against such tactics in the future. (vol. #12)

A few days after this (Aug. 25th), he finds he was mistaken in the above, and attempts to rectify any wrong done in a letter to Major Smith Standsbury, Comdg. Richmond Arsenal;

................................."In returning to you a letter which I supposed had been missent, and which referred to contractors paying higher wages and thus inducing hands to leave Govt. employ, I made a remark concerning Messrs. Spiller & Burr here having done likewise. After investigation, I find I was misinformed, and now wish to correct the statements as I am satisfied they are not doing as several parties report to me they were. It seems to have been a trick connived at by the mechanics to get higher wages in anticipation of a strike.

Hoping that my remark may not have done the party interested any injury." (vol. #12)

Meantime, the trickle of revolvers continued at the rate of 50 or 75 per month, and while most were sent to the Army of Tenn., a portion found their way to the Army of Northern Virginia. On Sept. 18th, the Atlanta Arsenal records receipt of 167 Spiller & Burr revolvers. Eleven days later the Arsenal shipped "One box Navy pistols to Col. Gorgas in Richmond." As there was never any substantial balance of revolvers kept on hand at the Arsenal, it seems reasonable to suppose this "one box of navy pistols" was a portion of the 167 revolvers received on the 18th. Revolvers were customarily shipped 50 to a box.

Oct. 16, 1863; "Col. James H. Burton having returned to Macon, the command of this Armory is resumed by him." (Order #22, vol. #49) This date marks Burton's return from Europe where he had purchased as much machinery as the shaky finances of the C. S. Govt. would permit. A rather complete account of his activities while abroad will be found in the "Appendix" section of this book to those interested, in letters from Burton to Gorgas.

### SPILLER & BURR PLANT PURCHASED BY THE C. S. GOVT.

By this time, Spiller & Burr were so far behind in the terms of their contract that any continuance in the hands of these private contractors was out of the question. Burton made immediate plans for its purchase by the Govt.

The purchase date was Jan. 9, 1864, but the actual payment to the firm was not made until Feb. 29, by Lt. Charles Selden, Jr. for $125,000 (order #1 1st Quarter 1864, Ordnance Service, C. S. A.). Selden's receipt shows he received this sum from Col. J. H. Burton, as Supt. of Armories.

The records of the Atlanta Arsenal show that Burton gave Col. Wright a receipt for $190,804.00 for the "purchase of the machinery of Spiller & Burr." (vol. #41)

There is a vast difference between the $125,000 as noted by Lt. Selden and the sum of $190,804 as stated by Burton, but as the purchase price has no bearing on this story we will not delve into this any closer, and let the reader make of it what he may. Suffice it to say that after Jan. 9, 1864, the pistols of the Spiller & Burr design were manufactured by the C. S. Govt.

## SPILLER'S ACTIVITIES AFTER THE PURCHASE OF THE PISTOL FACTORY

After his pistol factory was purchased by the C. S. Govt., Spiller went to Augusta, Ga., in a connection with the Endor Iron Works of Chatham, N.C. It is said that his father-in-law had some interest in this concern. At any rate in Sept. 1864, we find Spiller advertising for negro hands, and an iron founder. To the first, "a liberal hire wage, and locality most secure in the Confederacy" was offered, and to the second, a salary of $450 per month was promised to a "first class man." Both of these ads were inserted by E. N. Spiller, but applied to the Endor Iron Works.

Mention has been made of a Reese H. Butler, General Manager of the Pistol Factory. After its sale, Butler went to Raleigh, N. C., and became connected with Heck Brodie & Co., who operated as the Raleigh Bayonet Factory.

In May 1864, Col. Burton upon being asked by Gorgas to recommend some person to fill the position of Master Armourer at the Columbia, S. C. Armory says;

............."It is difficult to find men now who are fully capable to fill the position of M. A. at an armory, and I scarcely know who to recommend for the position; but feel of all those who occur to my mind, I know of none more fit than Mr. Reese H. Butler, later manager to Spiller & Burr, and now at Raleigh, N. C., where I am told he is in some bayonet factory at that place.

Mr. Butler is a comparatively young man, but I know him to be a good machinist, a steady correct young man and persevering, and I think he would do as well in the position of M. A., as anyone I know of who is available at present. I endeavoured to induce him to follow the pistol factory when removed from Atlanta to this Armory, but I believe he objected to being placed in a position subordinate to my Master Machinist, which I could not avoid.

My opinion is that Mr. Butler would do very well in the position of M. A. at a small armory like that at Columbia, S. C."

In Atlanta, near the Pistol Factory was a building which formed a portion of the C. S. Arsenal. Note the map of the City of Atlanta, dated April 12, 1864 (after the removal of Spiller & Burr), which was prepared by our old friend, L. P. Grant, whom we note is now a Capt. of Engineers, C. S. A.

On this map, #9 is listed as "C. S. Machine Works and Armory." According to Reed's History of Atlanta, this spot prior to the war was occupied by one James L. Dunning, an ardent Unionist, and who made no attempt to hide his views. He there operated a large foundry, but flatly refused to take any Confederate contracts. He was forced to dispose of his business to parties loyal to the Govt., and after that, the foundry turned out large quantities of war material for the Southern Army. Mr. Dunning, although an old and highly respected citizen was not permitted to go his way in peace. Military authorities arrested him and kept him prisoner for weeks. He did not know whether he would be exiled, shot or lynched, but he never wavered. He frankly avowed Union sentiments, but said he would obey the laws, and take no active part in aiding the Federals.

Finding the prisoner would not be converted or bulldozed, he was released and permitted to remain in Atlanta, unmolested as far as his physical being was concerned, but a thousand eyes watched his every movement, and a thousand ears were on the alert for some indiscreet expression of opinion.

The works were destroyed at the time of Sherman's little visit in 1864, but after the war were resurrected by Reese H. Butler, and a J. H. Porter, and for some years it was operated as Porter & Butler.

The 1867 Atlanta dietory lists the firm as being located at the SW corner of King St. and Georgia Rwy. Mr. Edward N. Spiller is listed as an employee there.

## REMOVAL OF THE PISTOL FACTORY FROM ATLANTA TO MACON

The following letters covering the removal of the Pistol Factory after its purchase by the Govt. from Atlanta to Macon, are to be found in Vol. #41, and speak for themselves:
Burton to Hiram Herrington, Master Machinist, Macon Armory, No. 28, 1863;

"You will proceed to Atlanta, Ga., without unnecessary delay, and report to Col. M. H. Wright, Comdg. Post, and show him this letter. The object of your visit will be to inspect a lot of pistols ready for inspection at the factory of Messrs. Spiller & Burr. On the completion of the inspection of which you will return to this Armory and make a report to me thereon in writing."

Burton to Gorgas, Dec. 22.

"I have the honor to acknowledge the receipt of your letter of the 15th inst., instructing me to assign Mr. Herrington, M. A. at this Armory to the duty of inspecting pistols manufactured at Griswoldville, Columbus, Atlanta, and Greensboro, Ga. and to prepare gauges for the inspection of same.

The gauges for testing the bore of barrel, and chambers of cylinder will be put in hand at once, and completed as soon as possible. That for testing the 'base pin drift in cylinder' I can not prepare in the absence of information as to diameter, etc., which is not given in the copy of Major Frank F. Jone's report, extract of which you enclosed to me. As soon as this is received here, the gauges will be prepared.

With reference to the assignment of Mr. Herrington to inspecting duty as above, I beg to state that if this is done I shall lose his services entirely at this Armory, which I am unwilling to agree to. In view of the probable removal of Spiller & Burr's factory to this Armory, Mr. Herrington's entire services will be required at least a month or two in re-erecting and starting the machinery, and you will doubtless recollect my mentioning Mr. Herrington to you in this connection when I first proposed the removal. I, however, respectfully suggest that Mr. Herrington may control the inspection provided he remains here, and an assistant inspector is provided who can receive his instructions from Mr. Herrington. Mr. Copeland might be assigned to this duty with advantage to the service perhaps.

Mr. Herrington's services are too valuable to this Armory to be dispensed with, otherwise I should be pleased to agree to his assignment to other duty."

Once a course had been set, Burton was not one to let grass grow under his feet. Jan. 6, 1864 Herrington is ordered to

................................"proceed without delay to Atlanta, Ga. and superintend the packing and shipping of the machinery, fixtures, tools, etc. & etc., of the Spiller & Burr Pistol Factory, in accordance with detailed instructions issued herewith." (vol. #41)

The same day to expedite Herrington on his way we find a note to Major J. G. Michaeloffsky, AQM—Macon;

............................."Please furnish Mr. H. Herrington, the bearer, with transportation from Macon to Atlanta, Ga. He goes on business for this Armory." (vol. #41)

Burton even found time to write Gorgas this same date (Jan. 6th);

"I have today been notified by Col. M. H. Wright, Comdg. Arsenal & Post at Atlanta, that he has consummated the purchase on behalf of the Govt., of Spiller & Burr's Pistol Factory in Atlanta, and I propose going to that place tomorrow to take charge of it, and to direct its removal to this Armory.

In the meantime, I think it desirable that no time should be lost in making an effort to secure a supply of steel from England for cylinders and barrels. I therefore respectfully recommend that orders may be sent out to England for the following steel—viz: 8,000 lbs C. steel for pistol cylinders 1⅝ round; 8,000 lbs C. steel for pistol barrel 71/100 octagon, to be obtained in Sheffield of Thos. Firth & Sons, Norfolk Works, who are familiar with the description of steel required for this purpose." (vol. #41)

Well might Thos. Firth & Sons be familiar with the type of steel used in revolvers. They were one of Samuel Colt's suppliers.

We gather from the above, that while Burton thought twisted iron for cylinders was all right when the manufactory was operated by Spiller & Burr, a year's experience with this substitute had convinced him that iron was not altogether-desirable, and to be avoided if possible.

### BURTON AND HIS FIRE ENGINE

Somewhere along the line, Spiller had picked up a fire engine, and at the time of the sale of his factory wished to dispose of the engine. This information was passed on to Burton who appears intrigued with the idea of actually owning a fire engine. For one day (Jan. 11th) he seemed more concerned with acquiring the fire-fighting apparatus than he was over the purchase of the revolver factory. On this day he wired Herrington; "care of the Spiller & Burr Pistol Factory, Atlanta, Ga.; Say to Richards I will take the fire engine. Will write you tomorrow and send you the money." (vol. #41).

So excited was Burton that he did not wait until the following day to write Herrington;

................"Your letter of the 9th inst. in relation to the fire engine etc. is received, and I have telegraphed to you this morning in reply that I will take it at the price named viz: $5,000. Mr. Selden will remit today to Mr. Spiller the money by express together with duplicate vouchers for Mr. Richards to receipt on payment to him of the money. You will observe that the engine, reel, hose, etc. are enumerated separately on the vouchers. Please confer with Mr. Richards, and agree with him on his prices for each item and fill in the same on the vouchers so that the whole will amount to $5,000. You had better ship the engine etc. as soon as you can to Macon." (vol. #41)

## LETTERS IN CONNECTION WITH THE REMOVAL FROM ATLANTA TO MACON

Burton to Gorgas, Jan. 14th.

"Messrs Spiller & Burr had employed in their Pistol Factory at Atlanta, Ga. at the time it was sold to the Govt., a number of workmen who had been detailed from the Army of Tenn. Several of these are excellent workmen, and I greatly desire their transfer to this Armory along with the machinery, etc. from Spiller & Burr's factory which is now being shipped from Atlanta. I am sure that public interest will be most promoted by the transfer of these men to this Armory. At the present moment I am unable to name the men, but I respectfully request that you will endeavour to procure an order from the proper authorities by which such of these detailed workmen as I shall be able to specify in a few days, may be transferred to me." (vol. #41)

Burton to H. Herrington, Master Machinist, Jan. 15th.

"Your letter of the 13th inst. is just received. Send me the particulars of the company, regiment, etc. of each of the 6 detailed soldiers you name, and I will make application to Gen. Johnston for their transfer to the Macon Armory. I have already written to Richmond generally on the subject, but could not then specify individuals.

In regard to the detailed conscripts, make out a list of their names (of those who you think it desirable to transfer here) and hand it to Col. Wright. I have telegraphed to him that I desired to employ all whom you would recommend and specify.

Keep as strong a force employed in the work of taking down and packing the machinery as you can, as I am desirous of your getting through as soon as possible. You had better look a little ahead in the matter of cars, in order that they may be ready to load when you want them. I shall hope to receive an invoice of some of the machinery by the beginning of next week." (vol. #41)

Burton to Major John F. Andrews, Major & AA, Gen'l Comdg., Camp Randolph, Decatur, Ga. Jan. 21st;

"The transfer of the Pistol Factory formerly carried on by Messrs Spiller

& Burr in Atlanta, to this Armory for the purpose of continuing the manufacture of pistols for the Govt., renders the transfer of all the mechanical force formerly employed in the establishment to this Armory a matter of very great necessity. I therefore respectfully ask for the transfer to this Armory of the following named detailed conscripts: J. A. Tuttle, T. P. Smith, R. H. Alley, C. W. Badger, D. A. Jones, J. M. Lamb, J. L. Morris, J. H. Wright, J. E. Rankin, T. H. Horton, T. L. Thomas, W. O. Rankin, John Breen, J. A. Fuss, Wade Hall, W. C. Moore, C. J. Owens, R. D. Rude, and W. H. Bowles." (vol. #41)

Burton to Gorgas, Jan. 23rd:

"On the 14th inst., I had the honor of addressing a letter to you on the subject of the necessity of transferring to this Armory the greater number of the detailed soldiers who had been employed in the Pistol Factory of Spiller & Burr at Atlanta. I now beg to enclose herewith a list of the names of the detailed men who I desire to be transferred, together with the particulars of their companies and regiments etc. I respectfully request that you will apply for their transfer to this Armory as soon as possible as the machinery etc. from Spiller & Burr's factory is arriving daily here, and I wish to set it up again and put it to work with as little delay as possible." (vol. #41)

Burton to Major A. Rowland, Comdg. Camp Cooper, Macon, Ga. Jan 27th;
"I have the honor to return enclosed herewith, my applicant to Col. J. Gorgas, Chief of Ordnance, for the transfer of detailed soldiers lately employed in the factory of Spiller & Burr at Atlanta to this Armory. Also a detailed list of the names, etc., of the men referred to. I respectfully request that these detailed men may be ordered to report to me at this Armory with as little delay as possible." (vol. #41)

## THE REMOVAL TO MACON COMPLETED

It is 97 miles from Atlanta to Macon, and the removal of a complete factory from one point to the other in 1864 was no easy undertaking. It speaks well of both Burton and Herrington, his Master Machinist, that less than a month from the date of purchase, the entire plant had been removed to the Macon Armory. Now all that remained was to re-erect it, set it in motion, and to start producing the 15,000 revolvers.

Burton to Gorgas, Feb. 1st;

"I have the honor to acknowledge the receipt of your telegram of the 25th inst. with reference to the setting up of the machinery received at this Armory from Spiller & Burr Pistol Factory in Atlanta. All the machinery is now here, and every exertion is being made to re-erect it and put it to work at the earliest moment possible.

I have also to acknowledge the receipt of your telegram of the 29th Jan. concerning the transfer to this Armory of the detailed soldiers and conscripts lately employed by Spiller & Burr. The transfer of the detailed soldiers has been arranged by order of Gen. Johnston, and that of the conscripts by order of Col. M. H. Wright, and nearly all of the men have reported to me for duty, and are employed in the re-erection of the machinery." (vol. #41)

Burton to J. Fuss, Acting Master Armourer, Feb. 1st;

"Now that this Armory has been supplied with a good fire engine, etc., it is very desirable that a company should be formed and organized to man it, composed of the employees of the Armory. You are hereby fully authorized to request the men to effect such an organization and to afford them such aid as may be necessary to accomplish this object. I will be pleased to assist the organization to the fullest extent in my power, as I desire to see it in the most efficient condition for service at short notice. You will please convey the purport of this letter to the employees of this Armory." (vol. #41)

Burton to Major C. J. Harris, Comdg. Camp Instruction, State of Ga. Macon, Feb. 12th.

"The bearer Robert Scott, has been transferred to this Armory from the Pistol Factory, Atlanta, Ga., by order of Col. M. H. Wright, Comdg. Post etc. Atlanta. He states that he enlisted as a private in Co. #H, 3rd Va. Vols. early in the war for 12 months; after 5 months service, he was detailed for special duty by order of the Sec. of War, which duty he has been performing up to the time of his transfer to this Armory two weeks ago. He further states that at the expiration of his 12 months enlistment he did not re-enlist nor has he been enrolled as a conscript.

If you agree with me in the opinion that he is liable to enrollment, please enroll him and detail him with orders to report to me at this Armory for assignment to duty, as his services are very necessary in connection with the manufacture of pistols about being commenced." (vol. #41)

### THE MACHINERY IS SET UP. BURTON IS STILL ANXIOUS TO GET STEEL

Burton to Gorgas, Feb. 21st.

"If there are any means by which you can expedite the delivery of the 3 tons of steel for pistol barrels and cylinders expected from Bermuda, please bring them to bear, as I am now ready to make use of it, if it were here.

The machinery is all set up ready for work again in this Armory, and on yesterday it was put in motion, and if the material in question comes to hand promptly, the manufacture of pistols can be pushed on vigorously.

I dislike to make use of iron for cylinders for reasons known to you, but unless the steel arrives soon, I shall be compelled to make use of iron to keep the workmen employed. If the steel arrives safely in port, please give such instructions as will prevent its being sent to any place other than Macon." (vol. #41)

Poor Burton, neither the records nor the pistols themselves give any indication that he ever received his much needed steel. On March 31st, he writes Gorgas disclosing a portion of his troubles in trying to substitute iron for steel;

"I have the honor to acknowledge the receipt of 2 requisitions for revolving pistols for the Army of Tenn. (200 and 724 respectively) approved by you. They will be supplied as soon as possible, but I beg to inform you that much delay and loss of time and labor results from the use of iron for cylinders. The last powder proof of cylinders resulted in the bursting of 18 out of 32 proved.

You will thus perceive that it will be some time before the above requisition can be filled unless steel is received for cylinders as expected from Bermuda. The proof test is not too great in my opinion and it is better that the cylinders should burst here than after issue. Can you hurry on the steel for cylinders?" (vol. #41)

Vol. #57, consists of a foreman's rough note book written in pencil, showing various workers' names, and the various items they made in connection with "The Pistol Factory, Macon Armory." This volume also shows the testing of iron cylinders during the 4 month period of March 1864 thru June. The results are rather amazing:

| Month | Cylinders proved | Cylinders burst | Cylinders cracked |
|-------|------------------|-----------------|-------------------|
| March | 151 | 46 | 71 |
| April | 279 | 12 | 38 |
| May | 274 | 27 | 74 |
| June | 416 | 78 | 137 |

No wonder Burton was so anxious to lay his hands on some steel!

The above table may give us the answer as to why so many Spiller & Burr revolvers are found with cylinders bearing no serial number. The serial was probably stamped on all parts at the time of assembly. Later the gun was "proved." If the cylinder burst under this proof, then a new cylinder was inserted. The new cylinder of course did not bear the serial number.

It seems odd however that in assembling the arm they did not use cylinders that had already been proved.

## LETTERS DEALING WITH THE MANUFACTURE OF PISTOLS AT MACON—1864
### (vol. #31)

Burton to Gorgas, Feb. 15, 1864;

"The enclosures herewith have reference to the case of one of the men transferred to this Armory from Spiller & Burr's Pistol Factory, Atlanta, Ga. Please have them examined and if possible obtain authority to have him enrolled and detailed as a conscript in which character I desire to deal with him for the reason I wish to appoint him to the position of foreman of the Pistol Factory now under my charge and as it is highly essential that he should be satisfied with his position I think this course I suggest will best accomplish the desired object. Scott states that he has never drawn any pay nor has he ever had [writing illegible]. He further states that his Regt. has been disbanded [3rd Va. Vols.]. If you obtain authority for his enrollment, please forward it to me as soon as you conveniently can."

Burton to Major Frank F. Jones, Inspector of Small-Arms, Ordnance Office, Richmond, Va. Feb. 16, 1864;

"Enclosed herewith I send you a copy of System of Inspection approved by the Chief of Ordnance for pistols manufactured by Spiller & Burr. Also the drawing of the press for extracting oil from lard such as is used at this Armory, and which I trust will be sufficiently plain for your purpose."

Oil from lard!

Burton to Gorgas, Feb. 18, 1864. Burton respectfully requests "a supply of small files are necessary to the manufacture of pistols" and wants half round

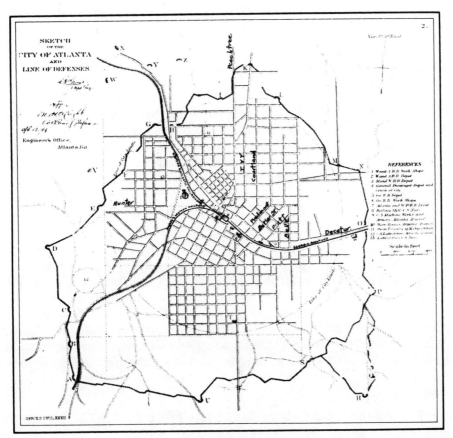

Map of Wartime Atlanta, Ga.

bastard 5 x 6 in., half round smooth 4, 5 x 6. Hand bastard 5 x 6, hand smooth 6 in., flat smooth 5 x 10, square bastard 6 in., 3 square bastard 6, 8 x 10, 3 square smooth 6 x 8, warding 4 x 5, round smooth 6-8-10 x 12. Wants as many as can be spared of the above, and also a general supply of others such as will be useful in an armory.

Evidently he was not very hopeful in getting the above order filled, as on the same date, he wrote a similar letter to Col. W. G. Rains, Comdg. Govt. Works, Augusta, Ga.

Burton to Fuss, AMA, Feb. 18, 1864;

.............................................. "Private John Cole, Col. A, 66 Ga. Regt., has been detailed from the Army of Tenn., by order of Gen. Johnston, for special duty at this Armory. Please assign him to duty."

### CRUCIBLES FOR PISTOLS

Burton to Gorgas, Feb. 26, 1864;

............................................"A supply of crucibles are required for the service of this Armory in connection with the manufacture of pistols. As I do not know the name or address of the maker in Petersburg or elsewhere, I respectfully request that you will order the Ordnance Storekeeper at Richmond to purchase the following and turn them over with as little delay as possible, to this Armory: 100 black lead crucibles #30, 25 same #40, 25 same #20."

## FOREMAN PISTOL FACTORY

Burton to J. Fuss AMA, March 5, 1864;

...................................."Mr. R. G. Scott, foreman of the Pistol Mfg. Dept. having been transferred to the Navy Dept. by order of the Secretary of War, it is necessary to fill the position thus vacated. Mr. P. S. Rodgers, machinist, is selected to fill the above place, and you will assign him to that duty on probation, commencing on Monday next 7th inst. Should he give satisfaction after the lapse of a reasonable time, he will be regularly appointed."

## PISTOL BARREL RIFLING

Burton to Gorgas, March 10, 1864;

................................"I have the honor to request the detail of Private Andrew I. Youngblood, Co. #H, 12th Ga. Regt. for duty at this Armory. Youngblood is a skilled mechanic, and is now employed at this Armory in rifling pistol barrels. His services are of the utmost importance in connection with the Pistol Factory. He was detailed Nov. 10th, 1862 under special order #263, A.I. Gen'l Office, and his detail having since expired, the enrolling officer has rec'd orders to send him to his regiment. At my request, Youngblood has received a 15 days furlough until I can hear from this application."

## READY TO RESUME MANUFACTURE OF PISTOLS

Burton to Gorgas, March 18, 1864;

................................."Annexed hereto is a list of materials required at this Armory which can not be obtained in this market. The materials are for the mfgr of pistols, and can not be dispensed with. The iron wire I have reason to believe can be made in Richmond, and if this be the case, I respectfully request that you will order the wire to be drawn and forwarded to this Armory as soon as possible. The sheet iron, steel wire, cooper tin and zinc, you will know best from what source to supply them and I also respectfully request that you will order some sent to this Armory with as little delay as possible. I am ready to go ahead with the mfgr. of pistols regularly after this month, but must have the necessary materials.
85 lbs iron wire, #6 wire gauge, 35 lbs same, #10, 35 lbs same, #4, 10 lbs same #19, 10 lbs same #15, 20 lbs sheet iron #16, 5 lbs same steeliron #20, 5 lbs same #10, 10 lbs same #16, 7000 lbs ingot copper. 1000 lbs pig tin, 2000 lbs zinc or speltre (?) cake. The iron wire may be mfgrd and supplied in straight lengths of about 6 feet in length done up in bundles."

## CANNON & STILLS INTO REVOLVER FRAMES

One can not help but wonder just how much hope Burton actually had in ordering such Materials? From the story of the Griswold & Gunnison revolvers, we saw how a portion of their brass framed revolvers were cast from church bells. Burton did them one better; May 17, 1864, Burton to Lt. Col. E. A. Delagnel, Comdg. C.S. Arsenal, Columbus, Ga.,

"On the 26th of March I was informed by the Ordnance Bureau that an old bronz gun would be turned over to this Armory from the Columbus Arsenal to be cut up and cast into pistol work. The gun has not yet been received and the material is much needed."

.........Again, Burton to Dr. T. E. Smith, Americus, Ga., June 29, 1864,

"In reply to yours of the 28th inst. I will pay you at the rate of $4.00 per lb. for the melting up with brass castings."

Maybe it is Dr. Smith's still, which gives such a red cast to some of the revolver frames made at Macon.

## BURTON PAINTS HIS FIRE ENGINE

Burton to Gorgas, March 22, 1864;

................................"Having secured for the protection of the bldgs of this Armory, and those of the other ordnance establishments in this city an excellent fire engine and apparatus, the workmen employed this Armory have organized themselves into an efficient fire company and take much interest in the subject. I am desirous of keeping active this interest as far as possible for reasons which will be obvious, and as one means of accomplishing this end, I desire that the fire apparatus should be newly painted, gilted etc., which can be done by the workmen employed by this Armory. I have hesitated however to purchase the necessary gold leaf without special authority which I respectfully request you will grant. The cost will be about $200."

## CRUCIBLES

Burton to Col. W. LeRoy Broun, Comdg. Arsenal, Richmond, Va. April 11, 1864;

........."On the 12th March, a case containing 18 crucibles was forwarded by special messenger to this Armory from the Ordnance Stores, Richmond. The crucibles appear to have been packed in wet sawdust or shavings, and the result is that 7 of them crumbled to pieces.
I call your attention to this subject with a view to the better packing of the 100 yet to be forwarded.
The messenger assures me that the case did not get wet whilst in transit."

Apparently the crucibles did not arrive from Richmond, as on May 18, 1864 Burton was forced to write to W. J. McElroy, noted sword maker of Macon,

........."You will oblige me by loaning to this Armory half a dozen crucibles for brass melting, of such size as you can best spare. I will return them as soon as an expected supply is received from Richmond."

Based on such a contingency, it is extremely doubtful that McElroy ever got his crucibles returned!

## PISTOLS MFGRY NOT PROGRESSING RAPIDLY

Burton to J. Fuss, AMA, April 23, 1864, "The mfgr of pistols does not seem to progress as rapidly as could be desired, under the pressure of this class of arm in the army. Please confer with Mr. Herrington, M. M. fully on the subject, and report to me what is necessary to be provided in order to increase the product to the maximum extent of the capacity of this machinery."

The next letter from Burton to Gorgas, dated April 25, 1864, reveals the Pistol Factory to have only one rifling machine, which may in part be the answer to the reason why things were not progressing as rapidly as could be desired.

"On the 10th of March, I addressed a letter to you requesting the renewal of the detail of private Andrew I. Youngblood, Co. H, 12th Ga. Regt. Vols., who is a machinist and has been employed at this Armory for the past year and a half. Since the removal of the Pistol Factory to this Armory, Youngblood has been employed at the work of rifling pistol barrels, on the only machine provided for that purpose. He has become very proficient at this special and particular work—the result of constant practice and painstaking. This morning I have been ordered by the commd of the Post to return Youngblood to his command by order of the Secretary of War. I have complied with the order but feel it my duty to inform you that the loss of Youngblood's services will greatly retard the mfgr of pistols at this Armory in as much as it will be necessary to instruct and educate another man to rifle pistol barrels to the degree of perfection attained by Youngblood. So highly do I estimate the value of his services at this particular work that I again request he may be detailed for special duty at this Armory as before."

### BURTON STILL NEEDS FILES FOR PISTOLS

Burton to Lt. Col. I. L. White, Comdg. C. S. Arsenal, Selma, Ala., May 3, 1864,

.........."I am informed that you have control of the Govt. portion of the cargo of the steamer 'Denbigh' lately arrived at Mobile, I notice in the invoice a lot of files assorted. I am much in want of files of all descriptions of 8 inches and under in length. Should you be able to spare any such, I beg to request that you will turn them over to this Armory, and forward them by express. The files are wanted in connection with the mfgr. of pistols."

### ANOTHER RIFLING MACHINE PLANNED

Burton to Fuss, AMA, May 16, 1864,

"It will be necessary to construct an additional machine for rifling pistol barrels with as little delay as possible. You will therefore give instructions accordingly to the Master Machinist in order that the patterns may be put in order and sent to the foundry as soon as possible and the other parts of the work put in hand."

### LACK OF MATERIALS—AND NOW THE LACK OF MONEY

Burton to Gorgas, May 13, 1864,

..............................."I beg to call your attention to the question of funds for the service of this Armory. No funds have yet been received on my estimate for the present quarter and the workmen are clamorous for the payment of the wages due to them. Some of them are represented as suffering for the necessities of life. I have made every effort to borrow funds for the payment of the April payrolls but neither the Banks or the Govt. Depository have any funds in the new currency. Unless funds are promptly supplied to pay workmen at the regular stated times, trouble will result. At the present time much discontent prevails amongst them and much time is being lost. If you can do anything to remove the difficulty I respectfully urge you to take action at once."

Ten days later Burton was still without money, and on May 24th wired his chief, "I am still without funds to pay wages. Cannot go on much longer without them. Please telegraph me when I shall receive them."

After his first receipt of crucibles from Richmond, portion of which were broken due to faulty packing, there is no indication that any more were received from this source, and on June 6th, Burton was forced to write to Capt. C. C. McPhail, Comdg., C.S. Armory, Columbia, S. C.,

............................................"Yours of the 23rd ult. has been received and I have to thank you for the 8 crucibles you have turned over to this Armory. They—nor the lightning rods have yet come to hand and I apprehend that they will need some special looking after. I am greatly in need of the crucibles, and in fact, am at a standstill in the brass foundry. Can you conveniently inquire after them? ......"

## MACON IS THREATENED

The summer of 1864 brought seige to Atlanta, Ga. Most Southerners were of the opinion that Sherman could just as easily conquer the Rock of Gibraltar, but even so, throughout the Confederacy there prevailed the uneasy feeling it was just possible that the heartland of the South faced some danger. Col. M. H. Wright, Comdg. Arsenal & Post, Atlanta, was definitely disturbed. In his correspondence with Gorgas he repeatedly expresses the opinion that he does not think Atlanta, in any real danger, but never the less, being a man of prudence, suggests that stores and equipment be forwarded from Atlanta to Macon and Augusta, "just in case," and on May 25th, supplies at the Arsenal began quietly leaving the city. Wright showing his usual judgment says that he "will not ship all at once with a view not to create any uneasiness with the people here."

Atlanta was some 100 miles from Macon, and there was hardly any likelihood of Atlanta being taken by the Yankees, and even if taken, certainly they would never get to Macon, but never the less Burton thought it would be a good idea to offer something in the way of a defense to his beloved Armory. On June 14th, he writes Capt. W. L. Reid, Comdg. Co. A, Armory Guard,

......."I refer you to the order published this day in relation to the organization of the military company by employees of this Armory on the 30th May ult. I desire that the drill may be commenced as soon as possible. You will therefore issue an order for drill on Wed. and Sats. of each week at 4 PM until further orders. The drill will take place at the temporary works—arms and equipment will be supplied as soon as I can obtain them.

You will issue an order requiring your men to assemble at the temporary works of the Armory promptly and without special orders in case of the approach of the enemy.

You will furnish me with the names of all who enroll themselves in your company whose names have not already been reported. All your reports will be made to me direct in writing.

You will call upon me for any assistance necessary to the efficiency of your command, and it will be cheerfully rendered if in my power.

You will communicate freely with me in this connection as I desire that the organization shall be attractive to the members, and creditable to all concerned."

## DESERTER TROUBLE

Burton to Major A. M. Rowland, Comdg. Camp of Instruction, No. 1, Macon, Ga., June 24, 1864,

........................"With reference to private J. S. Peck, now under arrest and confined at my request, for repeated and long continued absence from duty at this Armory. I think the good of the service requires that he should be made an example of. His absence of late is the result of an intention to join the Navy which intention I am of the opinion he should not be permitted to carry into effect. Some months ago, he and another deserted from this Armory, and were absent for nearly one month; common report said that they had gone to Florida with the intention of going over to the enemy. Difficulties, however, doubtless presenting themselves, they both returned and begged to be reinstated which in view of their usefulness, I agreed to do, on their promise that they would not give cause for complaint again. His companion has since deserted again, and not been heard of for nearly two months. Peck came from the North (born in Danbury, Conn.), his companion from Baltimore.

Putting all these circumstances together I think you will agree with me that the case of Peck demands special action. I propose therefore that he be assigned to the 4th Regt. Ga. Vols., and sent to the front without delay. I suggest this course as being preferable to awaiting the slow action of a court martial, to which I will otherwise subject him. You will oblige me by informing me whether or not you will adopt my suggestion in order that I may know what course to take."

## REPAIR OF ARMS AT THE ARMORY

Up to this time, the C. S. Armory at Macon, had been engaged in erecting a permanent establishment in the way of buildings etc. (Full reports of the Armory activities will be found under the "Appendix" section of this book). Until taking over the Spiller & Burr pistol factory, no arms had been made or repaired, and the only activity along this line was the supplying of stocks to Richmond for the rifles, muskets, and carbines. The action taking place in front of Atlanta called for emergency measures. Anything that could shoot or that could be made to shoot, was necessary at the front, to shore up Atlanta, which as the Arsenal records show was falling—falling.

July 7th, Wright reports that all machinery in the Atlanta Arsenal is down, and ready for shipment. (vol. 16)

July 8th, Wright reports all machinery and stores have been shipped to Augusta, except the percussion cap machinery which is being shipped to Macon. (ibid)

July 10th, Wright wires Macon, Columbus, Ga., Augusta, Ga., asking how many smooth bore muskets they have on hand. (ibid)

Vol. #16 is a huge ledger containing press-copy of letters sent by the Atlanta Arsenal. The last written page of this record is in lead pencil, and was evidently written in the heat of battle, by someone who had a sense of the ridiculous. It shows how garbled are the accounts of battle when they finally reach the backlines:

" 'Send me a saddle', a general engagement has taken place, Yankees defeated and in retreat.

'Send me a (?)', enemy falling back, no prospect of a fight unless we catch them.

'Send me a bridle', general engagement, we are whipped and are retreating.

'Send me a cartridge box', enemy moving towards the right, a fight expected.

'Send me a cap pouch', enemy moving towards the left.

'Send me a pair of stirrups', general engagement going on without material advantage to either party.

'Send me a haversack', general engagement has taken place, no advantage to either side."

Equally incongruous is a pencil notation found on the back of the front cover of the "Foreman's Time Book, Macon Armory 1863-64." (vol. 48)

"When you perhaps in after years,
May turn these pages o'er and o'er,
And read the names of friends so dear,
Should mine attract thy pensive eye,
Oh cast a lingering thought on me,
And remember, though I'm far away,
I'll ever love to think of thee!"

To continue from Vol. 31. We were saying that under ordinary circumstances, arms were repaired at the Macon Arsenal, which was under the command of Lt. Col. Cuyler, and not at the Macon Armory, Lt. Col. Burton Comdg. These, however, were not ordinary circumstances.

Burton to Cuyler, July 23,

........................"Colonel, I am ready to receive arms for repairs as soon as you send them to me. If you will send say a thousand or two today, I will employ my force on them tomorrow."

The following day, Burton, issues instructions to his Acting Master Armorer, J. Fuss, as regards the repairing of arms. Burton stresses the need of arms, and tells Fuss that no time is to be wasted in any attempt to "pretty them," "In repairing arms you will give instructions to the foreman that no labor is to be expended uselessly on 'appearance.' Special attention will be paid to the interior of the barrel to secure its cleanliness, the vent cleared and the lock put in such order as will secure its efficient action. All that is required is that the arms shall be made *serviceable*. You will direct your special attention to this subject in order that the greatest possible number of arms may be made available in the shortest time consistent with the requirements of the case."

Three days later, July 27th, Burton had repaired 200 muskets, and on this date wrote Major J. G. Michailoffsky, AQMM,

.............................."I have now 200 muskets ready to turn over to Col. Cuyler's store. Please send teams for them as soon as possible, and in the future send teams every morning until further notice for transportation of same number or arms. They will be packed in boxes."

Evidently some of these repaired arms were not all that were to be desired, and were cause of complaint from Cuyler, for on the 28th, Burton has this to say to Fuss;

............"Col. Cuyler informs me that several muskets repaired at this Armory have been returned to his store without cones and ramrods. You will please assign some competent person the duty of finally inspecting all repaired muskets in order to prevent a like omission in the future, and report to me the person you assign to this duty. All arm's chests sent to this Armory must be kept dry. I notice a number outside the shop which have been exposed to the rain this morning. Arms must not be packed in them until they are quite dry again."

## MACON PLANS HER DEFENSE

July 1864, and the encircling Federal forces were closing in on Macon, as well as Atlanta. July 9th, and Burton writes to our old friend Lt. L. P. Grant, Engineering Dept., Macon:

......................."Lieut. This will be handed to you by Mr. Augusta Schwaab, C. E., and Architect to this Armory, whose services I understand you will be glad to avail yourself of in locating the proposed defenses around Macon. I propose to lend him to you, conditioned that he shall continue to direct occasionally, the work he has charge of at this Armory, to do which he must visit the works from time to time. He will be borne upon my pay roll as heretofore, and you will please notify me as soon as you no longer may require his services. You will find in Mr. Schwaab an intelligent and able assistant."

With the expectation of action, more time was spent in drilling the Armory Guard,

........"Capt. W. L. Reid, Comdg., Co. A. Armory Guards, July 26, 1864. Capt., Tomorrow and thereafter your company will drill at 1 P. M., and 7:40 P. M. twice a week as heretofore until further notice. A quarter of a day will be allowed each member who attends drill. Arms will be issued tomorrow."

## ACTION AT MACON

Three days after this notice, the Armory Guards saw their first action. We are fortunate in having a complete description of this in a letter from Burton to Gorgas, dated Aug. 2nd.;

"I have the honor to report that the operations of this Armory were suspended from Friday evening 29th July ult., until this morning in consequence of the military company composed of the employees of this Armory being called out for the defense of this city against a large raiding party from Sherman's army, under the command of Maj. Gen. Stoneman whose object was the capture of this city and the release of the Federal Officers imprisoned here. The raiding party appeared before Macon early on the morning of Saturday 30th July, having with them two pieces of light artillery (9 pd. rodman guns) throwing Hotchkiss shells, and the action commenced on the high ground a short distance back from the Ocmulgee River, and opposite to this city, but within artillery range, and lasted until about 2 P. M. when in consequence of the gallant resistance offered by our forces, and also probably obtaining information of a force of our cavalry, being in their rear in pursuit of them, they retreated in the direction in which they came. They were completely foiled in their object in so far as it embraced the city and the Federal prisoners confined here; but they succeeded in damaging the Georgia Central Rwy., for a distance of 35 miles from Macon, burning several bridges, depots, trains of cars, etc. & etc., which will suspend communication with Augusta and other points East by R. R. for two or three weeks. Communication by telegraph is already resumed. Several shells were thrown by the enemy into this city, and the loss on our side during the action will amount to about 15 killed and 50 wounded. The loss on the enemy's side is not yet satisfactorily known.

In his retreat, Sherman and his command encountered about 14 miles north of Macon, a force of our cavalry under Gen. Iverson, a sharp contest ensued, and after inflicting considerable loss on the enemy, Stoneman, and his command surrendered unconditionally, and he and his officers, about 40 in number are now confined in the military prison here. About 500 of his men have already passed thru en-route to Andersonville, and it is expected that many more will be captured as it is believed that the greater part of the enemy's force dispersed into the woods at the time of the surrender.

I feel great pleasure in stating that all of my men who were physically able, responded promptly and with alacrity to the call. The order for them was received about 4.45 PM whilst at work and in a few minutes thereafter, they were mustered in the Armory enclosure, the roll called—the Capt. Comdg. received his orders and the company marched down to the Arsenal to be supplied with infantry accoutrements and ammunition—after which the Battalion composed of this company and the two infantry companies from the Macon Arsenal under the command of Lt. Col. J. W. Mallet, Supt of Laboratories, marched at once to the expected scene of action, and were on the spot before sundown, remaining in the field until Monday morning, Aug. 1st, at which time they were mustered out of service. No person connected with this Armory received injury during the action and every one performed his duty manfully and to my entire satisfaction. The balance of the day, on yesterday was given to the men to allow them to rest, and today the usual operations of this establishment are resumed (repairing arms)."

### ARMORY GUARD ARMED WITH SMOOTH BORE MUSKETS

A letter from Burton to Gorgas on Aug. 8th, indicates that his company had fought the action at Macon with smooth bore muskets, which weapon the good colonel evidently had no high regard.

"The military company composed of the employees of this Armory are at present armed with altered smooth bored muskets, cal. 69. The members take much interest in the organization and have requested to be supplied with rifled arms.

In the late brush with the enemy before this city, this company with others, felt the want of rifled arms to compete with those of the enemy. Most of the members of this company understand the use of rifled arms well, and many have been in active service in the regular army. Looking to the probability of repeated raids in this direction by the enemy, I am quite of the opinion that it would be a wise measure to arm this company with better arms than they have at the present. I am now repairing a large number of Austrian and some Enfield rifles from which the required arms (about 100) could be supplied. I therefore respectfully request that you will authorize the issue of 100 Austrian rifles to me by Col. Cuyler in exchange for the arms I now have."

Having had a taste of battle, the "guards" evidently regarded themselves as old soldiers, and wanted to be equipped as such.

Burton to Messrs. Barnett Micon & Co., Tallassee, Ala., Aug. 17th, 1864:

"The Military company composed of the employees of this Armory desire to uniform themselves if the clothing can be purchased at Govt. prices. Please inform me if you will furnish sufficient grey cloth to uniform 120 men at the Govt. price, and enclose to me a sample, stating the price per yard. Your early attention will oblige."

Burton to Lt. Col. R. M. Cuyler, Comdg., C. S. Arsenal, Macon, Ga., Aug. 29.

"I respectfully request that you will turn over to this Armory for the equipment etc., of the military company composed of its employees, the following articles and ammunition viz: 100 cap boxes, 43 cartridge boxes and belts, 100 bayonet scabbards, 44 canteens, 4,000 cartridges for Austrian rifle cal. 54 with requisite percussion caps. I will be glad if you will turn over the ammunition at once if possible, also the cartridge boxes and canteens. I am anxious to complete the equipment of the entire company as soon as possible, as its services may be required at any moment."

That the Armory was in daily expectation of a raid can easily be seen from Order #35 of the Macon Armory, dated Aug. 14,

..................................................."Information of a raid by the enemy in strong force on some point in Middle Georgia has just been received by the Supt. All members of Co. #A Armory Guards, are therefore hereby ordered to hold themselves in readiness in every respect for active duty. The firing of 3 guns in rapid succession at the Arsenal will be the signal for the company to muster at the Pistol Factory with all possible dispatch. It is hoped and expected that all members able to do duty will rally at the signal." (vol. 49)

## COMING EVENTS CAST THEIR SHADOW

Atlanta was occupied by the Union General Tecumseh Sherman on Sept. 3, 1864, but long before its actual fall, the "shadow of coming events" became apparent to the nearby Macon Armory.

On Sept. 2, Burton ordered his Acting Master Armorer Fuss to take down and pack all stocking machinery, and the machinery for pistol manufacturing. On the 5th of the same month Burton writes Gorgas that he is sending this machinery to Savannah for storage, but alas, he could find no transportation. Upon receipt of the above, Gorgas wired not to move the pistol machinery, but to send the stocking machinery to Columbia, S. C.

Vol. #29 of the Captured Confederate Ordnance Records, concerns itself with pressed copy letters which were sent from the Macon Armory from Aug. 29, 1864 to April 17, 1865.

Only a glance through this is needed to see the real state of affairs within the Confederacy. In it one finds a suggestion to remove the machinery from the Govt. Pistol Factory at Columbus, Ga. (formerly L. Haiman's & Bro.) to Columbia, S. C. and to discontinue the manufacture of pistols there and utilize the machinery for making rifles. There too is found a confession on Sept. 23rd to Major F. F. Jones, Comdg., C. S. Armory, Richmond, Va., that there were no more gun stocks available due to the "supply of blanks having been exhausted and no additional received despite all efforts. The machinery has been down for the last 4 months and packed ready for shipment."

On the same date in reply to a rather frantic request from F. W. Cook, head of Cook & Bros. rifle factory in Athens, Ga., as to where he should remove his machinery "in order to assure its safety," Burton is forced to say that he is unable to advise any such point. On 27 Sept. Burton writes Gorgas that Cook's Armory at Athens, has been closed for the past two months because the Govt. has failed to pay anything for work done since March 1st, last!

There is more, much more, in this volume referred to, but already the reader must have a picture of a mighty unstable business.

## *PISTOL MAKING RESUMED*

We digress. Let us return to the pistols, using the actual correspondence as nearly as possible to follow their operations.

Burton to Gorgas, Aug. 29,

............................"The set of stock machines at this armory has been, and is still idle, in consequence of the want of gun stocks in the rough, and there is no prospect of an early supply. As a measure of precaution would it not be well to have this machinery boxed and packed and removed to some point less exposed to the raids of the enemy? The necessary packing boxes have been made some time. Anticipating difficulties in securing RR transportation in an emergency, I have caused to be made several large flat-boats, 60 feet long, and 14 feet wide with a view to floating them loaded with property down the Ocmulgee River to some point in the State comparatively inaccessable to the enemy. A glance at the map of Ga. will inform you as to the safety and expediency of such a measure. In my humble opinion I think it the best course that can be pursued. If you approve this removal please telegraph me of same."

Burton to Fuss AMA, Sept. 2nd,

............................"Sir, You will make arrangements this afternoon for the taking down and packing in cases prepared for that purpose, all the stock machines, and the pistol machinery. The work of packing will be commenced on tomorrow morning, and the necessary force of carpenters and machinists will be detailed at that time. Let the force be ample for the prompt execution of the work."

Burton to Gorgas, Sept. 5th,

............................"I have been shown by Lt. Col. Cuyler, the telegram you sent him under date of 3rd inst., on the subject of the removal of surplus stores etc. In view of the fact of the evacuation of Atlanta, by Gen. Hood, and his army (which fact may not have been known to you at the time you dispatched to Col. Cuyler) and his retreat to a point on the Macon & Western Rwy., some 27 miles South of Atlanta, and acting upon the advice of Col. Mallet, and Cuyler, and Gen'l Cobb, Comdg. Ga. Reserve Forces, I have decided to pack up all the stock machinery and that of the Pistol Factory and ship it to Savannah, to be followed by other machinery and stores, should necessity prompt such a course. As neither the stock machinery or the pistol machinery is at present employed, no interruption to the work at hand (repair of arms) will result, at least for the present. Col. Mallet has just returned from Savannah, and reports that he has secured suitable store houses for the reception of all the machinery & stores likely to be sent from Macon Ordnance Establishments. Most of the stock & pistol machinery is already packed ready for shipment."

Telegraph from Burton to Gorgas, Sept. 8th,

.................................................."Pistol and stock machinery packed ready for shipment, but no transportation can be furnished at present. Shall I ship to Savannah when transportation is furnished? Prisoners are being moved from Andersonville."

## GORGAS SAYS NOT TO SHIP THE MACHINERY

Burton to Gorgas, Sept. 10th,

..............................."I have the honor to acknowledge the receipt this day of your telegram of the 9th, instructing me 'not to move the machinery.' Also of my letter to you of Aug. 29th, with your endorsement thereon in which you suggest the removal of the stock machinery to Columbia, S. C., and request me to express my views in relation thereto.
In compliance with the instructions as contained in the above telegram I shall not move any machinery etc. at present, and will await further developments of enemy's plans and developments. I will however allow the machinery to remain packed for the present. With reference to the removal of the stock machinery to Columbia, S. C., I think it is a good suggestion, and respectfully recommend that it be carried into effect at once, and set up there under the direction of Capt. McPhail. It will be necessary to provide a bldg. to receive it, which may be erected of wood, one story high. The engine at Columbia will afford the requisite power to drive this machinery in addition to the machinery now there, and the economy of transportation of stocks will be great in as much as the rough article must be supplied from that region."

Communications throughout the Confederacy were pretty well disrupted during this period. It was almost a month before Burton received a reply, as is shown by his letter of Oct. 6th to Gorgas;

................................."I have the honor to acknowledge the receipt of your telegram of the 23rd Sept. ulto., ordering stock machinery to be sent to Columbia, also your letter of the 27th Sept. instructing me to ship to Savannah as soon as possible 'all supplies, spare machinery and tools not required for immediate use,' also your letter of the 30th Sept., instructing me further with reference to the removal of machinery at this Armory, and finally your telegram of the 5th inst., instructing me to 'put the Pistol Factory in operation and push the work.' All of these orders and instructions have been, or are being complied with to the best of my ability.

The stock machinery is now being loaded on cars, and will leave here tonight or tomorrow morning. Surplus machinery here, and stores are now being packed and will be sent to Savannah. The pistol machinery will be re-erected at this Armory as soon as possible, and the mfgr. resumed and pushed on with all possible vigor."

This same date, Burton writes Fuss, his AMA to the effect that he has been notified by Lt. Col. R. M. Cuyler that no more arms will be sent to the Macon Armory for repairs; that Gorgas has ordered the re-erection of the pistol machinery, and to push on the mfgr. of pistols with all vigor possible. Fuss will therefore proceed at once to unpack and re-erect that machinery with all force available, and report as soon as operations can commence.

Being a man of some foresight, Burton adds that the packing boxes in which the pistol machinery is packed, had best be saved—they might have need for them again.

Two days later, Burton advises Fuss that he has instructed Lt. Selden, Military Storekeeper to retain only 2 months supply of material necessary for the mfgr. of pistols, and for other current requirements of the Armory. Says the work to be done will probably be limited to the mfgr. of pistols, and the erection of buildings for the permanent Armory.

Oct. 11th, Burton writes Fuss;

................................."I am in receipt of your report of the fact that the pistol machinery being ready for the resumption of the mfgr. of pistols. You will please issue instructions for the resumption of that work at once and for its vigorous prosecution. All work heretofore done by the piece will be discontinued, but prices will be raised in accordance with the late revision of wages etc., but not until I confer with you in relation thereto."

## MANUFACTURE OF PISTOLS RESUMED

On the 13th of Oct., Burton wires Gorgas that the pistol manufacturing machinery has all been re-erected and the manufacture of pistols resumed.

A "Summary Statement of work done at Macon Armory, Month of October 1864, Pistol Factory & Machine Dept." shows the following:

"50 revolving pistols completed. 460 Austrian rifles cleaned and repaired. 56 machines together with countershafts, hangers, etc., therefore comprising nearly all the machinery in machine shop have been taken down, boxed and packed for transportation to another point if necessary. All the machinery for the mfgr. of pistols has been unpacked, and re-erected, and the mfgr. of pistols resumed."

Well, at least 50 revolvers were finished during the month of October.

## CONFEDERATE THANKSGIVING DAYS

Nov. 16th, had been set aside by Pres. Jefferson Davis, as Thanksgiving. This day was to be spent fasting and praying. Those in charge of the Macon Govt. establishments were of the opinion the day could be better spent in making arms.

Burton to Cuyler, Comdg. Macon Arsenal.

............................................"In reply to yours of yesterdays date with reference to observing tomorrow as a day of fasting and prayer under the President's Proclamation. I beg to state that as yourself and Col. Mallet have decided to keep open your respective establishments I shall do the same, and issue an order notifying all my employees that my shops will not be closed, and that absentees without authority will be punished as provided for in Army Orders #37, current series."

It had been different the Thanksgiving of 1862, for then, Burton had issued an order somewhat more pleasant for those concerned;

..................................................................."Tomorrow (Thurs. Sept. 18th inst.) having been appointed by his excellency Pres. Davis as a day of Thanksgiving, in consideration of the recent triumphs of the Confederate arms—the Armory will be closed that day." (vol. #49)

Compared with 1862, a lot of things were different. The triumph of Confederate arms in 1864 were few and far between.

## THAT FELLOW SHERMAN AGAIN

The resumption of pistol making was of short duration. Enemy operations were such that all concerned were fearful of Macon's fall. Stores, machinery, equipment were all packed and shipped from the Armory, the Arsenal, and the Laboratory, to Savannah and to Columbia, S. C. It was to this latter point that a portion of the pistol factory was sent. Although Macon did not fall, as was expected, never the less for a period of about 3 weeks, no work of a productive nature was forthcoming from any of the above named establishments.

On Dec. 7th, Burton writes his Chief (Gorgas), outlining the events of the past few weeks, and also his plans for the future. Gorgas incidently, had just been promoted from the rank of colonel to that of Brig. General, and this is Burton's first letter to so address him.

"General:

I have the honor to report that in consequence of the advance of General Sherman and his army through Georgia, and in compliance with your instructions of the 30th Sept. ulto., with reference to the removal of the machinery and stores at this Armory, in the event of Macon being threatened by the enemy, and also acting under the advice of Major General Howell Cobb, Comdg. Milt. Dist. of Ga., with whom I conferred personally, the entire machinery at this Armory was taken down about the middle of last month, and packed in cases for transportation and the greater portion of it sent away as follows: All the machines in the charge of the MSK were sent to Savannah, Ga. Also the greater portion of the machines in the machine shop. The machinery for the manufacture of pistols I regarded as pertaining to the New Armory, and therefore decided to send it to Columbia, S. C., along with the foreign machinery and such as had been constructed at this Armory for the mfgr. of arms. Only about one half of the pistol machinery was shipped before the Central RR was occupied by the enemy, and also about two thirds of the foreign machinery, and about one half of the arms machinery constructed here.

No stores have been sent away and consequently all are still on hand here. So far as the machinery of machine shop is concerned, its absence will not materially affect active operations here as workmen have not been available to employ it for some time past. I have re-erected two or three lathes, drill planer, etc., etc., in the machine shop and fortunately had on hand a small portable steam engine of about 2 HP, which I have had placed in position in the shop and which furnishes power sufficient to drive about 60 ft. of shafting and the above mentioned tools. This enables me to make necessary repairs to machinery, etc.

With reference to future operations at this Armory, I respectfully recommend as follows:

*1st.*—The completion for occupancy at once of the wing of the new Armory bldg., now being roofed in and which will be completed with temporary roof of shingles in from 3 to 4 weeks from this time.

*2nd.*—The re-erection on the 2nd. floor of this wing of the pistol machinery.

*3rd.*—The re-erection on the 1st. floor of same wing of the machinery of machine shop for which purposes this portion of the bldg. was originally designed.

*4th.*—The erection of the steam engine and boiler lately employed at temporary works under a temporary shed in the yard adjoining the above specified wing for the purpose of driving machine shop and pistol factory.

*5th.*—The abandonment of the temporary works for mfgry. purposes and the concentration of all at the new works.

The advantages of this arrangement will be obvious. The appropriation of the bldgs. at the temporary works will be referred to in a future letter. I have already in anticipation of your approval of what I now recommend, commenced the re-erection of the steam engine and boiler at new works and am pushing on the bldg. to completion. In the meantime the armorers are employed in repairing arms, and in performing such work on parts of pistols as may be done without machinery.

Genl. Cobb is engaged at present in making arrangements for the establishment of a wagon train between Milledgeville and Mayfield—the present terminus of the Warrenton Branch of the Ga. RR (35 miles) which he expects to have in operation in two weeks' time. He kindly offers to transport the pistol machinery on his wagons which will be available for this purpose. The machinery is light, and can readily be thus transported.

I respectfully request your approval of my suggestions, as here-in-contained at the earliest moment possible in order that I may make all necessary arrangements with as little delay as possible.

P. S.—This will be sent by the line of couriers established by Genl. Cobb, via Milledgeville and Mayfield by which route a reply will soonest reach me. A telegram thru Col. Rains (Augusta, Ga. Arsenal) will be the shortest method of communication."

On the same date he writes Major J. T. Trezevant, Comdg. C. S. Arsenal, Columbia, S. C., to the effect that he would like the return of his machinery which had been sent there for temporary storage.

The "Summary of Work" for the month of November, shows 35 pistols made, assembled, and ready for proving. These having been made evidently prior to the shipping of the machinery in Nov.

In Dec. these 35 pistols were proved, and an additional 12 made up, probably from parts already on hand. As will be shown, these 47 revolvers were the last complete items turned out by the armory, for although Burton had written to Gorgas that he hoped to have his machinery back from Columbia, S. C., in two or three weeks, the fact is, it was never returned.

This same December, Fuss, AMA, writes Burton,

............................................. "If it is intended to resume the manufacturing of pistols at this Armory I would recommend proceeding with the casting of the lock frames as it would always be best to have that branch of the work in advance. Other employment can readily be furnished the founder when it may be desired to suspend casting." Fuss continues, "No report of operations for November was made because there was no work completed during the month. A few pistols—35 were assembled ready for proving, but as they are unfinished, it was deemed improper to report them."

A later report from the Armory, dated Dec. 31st, to Richmond, Va. as to work performed at the Armory, shows Fuss's suggestion had been carried out, as it lists 164 lock frames cast.

Vol. #30 is composed of copies of the letters sent by AMA Fuss of the Macon Armory. On Jan. 11, 1865, he informs Burton of the loss of 5 pistols:

"Sir:
 I have to report that 5 pistols have been stolen from the Armory since above the 20th of Nov. ult.

These pistols were a part of a small lot (11) of defective ones, retained because they were not deemed good enough to be put into service as they were, as some parts of all of them were defective. They were packed in one of the boxes made for packing pistols at the time the machinery and tools were packed to be sent away, and not opened until yesterday when the theft was discovered. The circumstances that led to the examination was the discovery of an attempt on the part of some one in the Armory to steal a vise on the evening of the 9th inst. About the time the bell was being rung on that evening a vise was thrown over the fence into the street in the rear of the Machine Shop. A gentleman who happened to be passing heard the vise fall and on examination found what it was and he came to the office at once and reported the fact to Mr. Stone, who was the only person remaining. Mr. Stone placed the guard so as to detect the thief, when he returned at night as it was expected he would, but unfortunately did not, so we have no clue as to who the offender may be. The box of pistols were in the assembling room, the top of the box was both screwed and nailed down, and Mr. Bass, who had the key of the room in his charge says he has never left the room for 15 minutes without locking the door and taking the key with him, so the inference is that the theft most probably was committed at night or on Sunday. I propose to charge some reliable man in the shop who may not be suspected on such duty with the duty of keeping watch over the movements of the men at such times as these depredations are most likely to be attempted, and I think it would be wise to do so, even if a small expense was incurred thereby."

His statement to Burton as to the work done in the Machine and Pistol Dept. for the month ending Jan. 31, 1865, is enlightening. Besides showing what was done, it also shows that no revolvers were completed, and that they were working on parts only.

"272 Austrian rifles, cleaned and repaired, cal. 54; 56 Austrian rifles cleaned and repaired, cal. 58; 153 smooth muskets cleaned and repaired, cal. 69; 51 rifled muskets, cal. 71; 10 rifled muskets, cal. 75; 9 smooth muskets, cal. 75; 25 carbines—various sizes; 1,191 lever catch studs, trimmed & filed; 342 main springs, filed and set; 116 lock frames cast; 280 lock frames drifted; 166 guards cast; 160 guards first milled; and 500 triggers filed."

Less than a month before the end of the war, we find the following letter from Fuss to Burton. This is dated March 21st, and clearly indicates that as of that date no revolvers were being turned out.

"Sir:
The following tools, and machines will be required for resuming the manufacture of pistols at this armory. Viz: 1 rifling machine, 2 barrel boring machines, 1 screw making machine, 2 lathes, 1st. drilling barrel, 1 cone drilling machine, 2 lathes, 1st. drilling cylinders, 1 cone tapping machine. Tools & holders for drilling & tapping lock frames. Tools & holders for milling sides of hammers. Tools & holders for milling edges of catch levers. Tools & holders for milling sides of triggers.

Temporary arrangements may be made for performing some of the operations by which some of the tools named above could be dispensed with for a time. The drilling cylinders for instance, Mr. Herrington thinks can be done on a drill press. The tapping cones, and milling and tapping screws may also be performed by hand, and means can be devised no doubt for doing other parts of the work in the same way, until the machines and tools essential for turning out the work with greater accuracy and speed can be made."

The same date, Fuss writes to H. Herrington, Master Machinist, and orders him to start to work on the machinery which will manufacture the muchly needed revolvers.

"Sir:
You will at once have the work commenced on the following tools. Viz: Tools & holders for drilling & tapping pistol lock frames, milling sides of hammers, milling edges of catch levers, and milling sides of triggers.

Please also to have the patterns for machines for drilling barrels and cylinders, for boring barrels—drilling and tapping cones, and screw making and slitting, put in order at once and send to the foundry with the number of castings required from each pattern. These machines will be fitted up as soon as possible, omitting all unnecessary finish by which time may be saved. You will also devise a simple machine for rifling barrels one at a time, each groove to be finished before another is commenced. The following machines imported from England and in the hands of the M. S. K., and if any of them can be made available in the Mfgr. of Pistols, you can order them out and adapt them to such work as will be found practicable, viz: 1 vertical drill, one spindle, 3 do. four spindles, 2 screw milling machines, sliding spindles with levers, 6 straight milling machines with upright stays.

In the meantime let such work as can be done on parts of pistols without present facilities and such means as can be temporarily adopted for promoting the work be resumed and put underway without delay."

Of general interest to collectors is a sword in the Battle Abbey, Richmond, Va., a part of the late Richard D. Steuart Collection, which was made in Macon, Ga., by W. J. McElroy Co. The blade is finely etched with acid, and contains the names of both W. J. McElroy, and that of H. Herrington, Master Machinist Macon Armory. This shows the close association of personnel at the various arms making plants in Macon.

The final "Summary of Work Done at the Macon Armory, in the Machine & Pistol Dept.," was for the month of March, and was submitted by Herrington on April 3rd., 1865.

"REPAIRED—the following with bayonets fitted: 116 smooth bore muskets, cal. 69; 17 Austrian rifles, cal. 54; 6 rifled muskets, cal. 71; 9 rifled muskets, cal. 58; the following without bayonets: 42 Mississippi Rifles, cal. 54; 10 rifled muskets, cal. 71; 16 rifled muskets, cal. 58; 89 smoothbore muskets, cal. 69; 42 musketoons, cal. 69; 70 carbines, cal. 58; 23 rifles (turned over in Febry. last), cal. 54."

"MANUFACTURED—970 guards cast; 409 Lever catch studs, filed; 1909 Lever catch studs, trimmed; 153 lock frames drifted; 529 guards first milled. Engine boiler and counter shafting for same completed and ready to start. Two new lathes for drilling barrels, two for drilling cylinders and two machines for boring barrels in progress, also one stock for drilling cylinders. Main lines of shafting in new shop put up and ready to run, also counter shafting for polishing machines."

And this is as far as the machinery ever got to its resumption of pistol making, as the war was over before the end of April.

## SUMMARY OF THE SPILLER & BURR REVOLVERS

Summarizing to this point, Spiller & Burr, in Nov. 1861 contracted to supply the C. S. Government with 15,000 revolvers to be delivered at the rate of 5,000 per year. The firm was organized in Richmond, Va., but in early summer 1862 removed to Atlanta, Ga. "Sample revolvers" were not forthcoming until Dec. 1862, and the actual working specimens did not appear until late spring 1863, and continued until the plant was purchased by the C. S. Government in Jan. 1864. At that time, the Pistol Factory was removed from Atlanta, to Macon, Ga., under the supervision of Col. J. H. Burton Comdg., Macon Armory. The year 1864 saw a number of revolvers of the Spiller & Burr pattern turned out at this Government establishment, with a continuation of the serial number as started by Spiller & Burr. The manufacture of these revolvers was halted in August and September due to enemy raids during which time the machinery was packed and ready to be sent to a point of safety at a moment's notice. Manufacture was resumed in Oct. and continued briefly until the middle of November, when enemy operations again caused the Armory to suspend work and remove in part to Columbia, S. C. In December 1864 a few pistols were assembled from parts. After this date none were made, although parts were turned out until the end of the war, and resumption of manufacture was planned and anticipated.

The question arises as to how many of these pistols were actually made by Spiller & Burr, and how many by the C. S. Government? We will attempt to answer this question a little later.

## MARKINGS ON REVOLVERS

Reference has already been made to the fact that some of these revolvers are marked "Spiller & Burr" on the barrel top while others are devoid of such stampings. "Firearms of the Confederacy" by Fuller and Steuart, has this theory to offer on page #281, as to why some are marked, and some unmarked; "Of these Confederate 'Whitney's' in museums and private collections, there is no record of a 'Spiller & Burr' (one so marked) with a serial above #600. It has been assumed that the unmarked revolvers were made by the Government after the purchase of the Spiller & Burr Plant in Feb. 1864, but this is only conjecture."

This theory is good only as long as actual serial numbers of known existing specimens are not examined side by side. Serial numbers of known specimens indicate the total number of revolvers by Spiller & Burr, plus those made by the Government falls under the 1500 mark.

## LIST OF SERIAL NUMBERS KNOWN TO AUTHOR

Let us examine the serial numbers of the guns known to the author. These have been collected over a period of years, and have been transcribed from one piece of paper to another so often that there is a distinct possibility of error in both the serial number itself and additional markings. However, such errors if they exist should be relatively small, and will not affect the over-all picture.

#72, unmarked except for serial.
#75, serial and firm name.
#77, serial and firm name.
#81, believed to contain only serial.
#86, serial and firm name.
#98, serial and firm name.
#101, markings unknown.
#104, serial and firm name (CS. on right side of frame).
#124, serial and CS on left side of frame.
#128, markings unknown.
#131, believed to have only serial.
#150, serial and firm name (C.S. on left side of frame).
#160, serial only.
#209, markings unknown.
#241, markings unknown.
#268, serial and firm name.
#272, markings unknown.
#319, serial and firm name.
#370, markings unknown.
#421, markings unknown.
#434, serial only (C.S. on left side of frame).
#490, serial and firm name.

#498, believed to contain only serial.
#535, serial and firm name.
#564, markings unknown.
#570, markings unknown.
#585, believed to contain only serial and CS on right side of frame.
#616, markings unknown.
#666, markings unknown.
#703, serial only.
#742, serial only, CS on left side of frame.
#763, serial only, CS on left side of frame.
#828, believed to contain only serial.
#855, serial only.
#903, serial only.
#905, serial and firm name.
#965, serial and firm name.
#969, serial and firm name.
#983, serial only (CS on right side of frame).
#988, markings unknown.
#1028, serial and firm name.
#1031, markings unknown.
#1058, markings unknown.
#1062, serial only, CS on left side of frame.
#1140, serial only, CS on left side of frame.
#1182, markings unknown.
#1214, markings unknown.
#1234, serial only, CS on right side of frame.

In addition to the above, two additional Spiller & Burr's are known. These bear no serial or firm name. One has already been described as possibly a "sample." The other, never seen by the author, is stated to be unmarked except for "CS" on the left side of the frame.

It might be well to state that on the above given serials, no attempt has been made to accurately establish which revolvers are stamped "CS" in addition to the serials. Those which are so noted were definitely established, and the statement "unmarked except for serial" does not mean that the gun might not be stamped "CS" in addition.

We have before us then, a list of 48 Spiller & Burr revolvers of which at least 4 are found carrying the firm name, and whose serials are above #600. Also to be noted are 8 below #600 which do not contain the firm name.

### WHICH REVOLVERS WERE MADE BY SPILLER & BURR, AND WHICH BY THE C. S. GOVT.?

We will soon establish that at the most, only 1451 revolvers could have been made through the combined efforts of Spiller & Burr, and the C. S. Armory Macon. Which then were made by Spiller & Burr, and which at the Armory?

Examining Vol. #1, of the Captured C. S. Ordnance Records, we find the "Monthly Consolidated Components and Pistols Fabricated at the C. S. Armory, Macon, Ga." This volume shows the assembly of pistols by the month from the time the revolver factory was purchased in Jan. 1864 thru Dec. of the same year. Remember now we have already established that no revolvers were made after 1864, parts only were fabricated after this date. In other words the

Barrel marking of a "Spiller & Burr." Note the tail of the final "R" in "Burr" was evidently broken on the die.

pistols manufactured during the year 1864 were the TOTAL which were made at the Armory, by the C. S. Govt. All previous to Jan. 1864 were made by Spiller & Burr.

Vol. #1 reflects the assembly of pistols for the year 1864 as follows: January-0, February-0, March-100, April-150, May-100, June-162, July-80, August & September-0 (during these months the machinery was crated and ready to be sent to Savannah), October-50, November-35, December-12.

This is a total of 689 revolvers which were actually turned out at Macon under Govt. manufacture.

In the Confederate Museum in Richmond, Va., is a brass-framed Whitney serial #763, which was the property of General Rains of the Augusta Arsenal. According to Gen. Rains, this pistol had been presented him "as a product of the Macon Armory." This proves that the manufactory under private management did not exceed 762 revolvers, for obviously otherwise, serial #763 could not have been presented to Gen. Rains "as a product of the Macon Armory."

Records show clearly that Col. Burton was an admirer of Gen. Rains, and sought his advice on several occasions through their mutual correspondence. This being so, it seems reasonable to suppose that Col. Burton in making a gift of one of his revolvers would have selected one of the first, if not the very first pistol which was turned out at the factory under his supervision. If this assumption be correct, then nearly 762 revolvers were made under private management, and 689 were made under Government supervision. At the most no more than 1451 revolvers could have been made, and if our assumption that Rains was given one of the first is incorrect, then this total will be less.

If our reasoning is correct, how can we account for the fact that some revolvers are found over the 762 serial are marked "Spiller & Burr"? One short sentence from Col. Burton to Col. Gorgas on Dec. 4th, 1862, gives us the complete answer to the question that has plagued historians for so long. On this date, Burton reported that he had inspected the Spiller & Burr factory at Atlanta, and found it almost ready to turn out the finished weapons, "—having on hand 2,650 barrels in the process of manufacture."

This one sentence gives us an insight hither-to-fore lacking, as it indicates very clearly that three months before the first revolver was turned out, they already had 1000 more barrels than were ever actually used!

Barrels of the Whitney model revolver are not an integral part of the frame. They are merely screwed into the frame, and were made from octagon rods of steel. The micrometercaliper is not a recent invention, and was in use at the time of the War Between the States. As a matter of fact, Burton brought one back with him from his trip to Europe, and offered to sell it to the C. S. Govt. Thus, the fact that the barrels were octagon in shape would not prevent their being stamped prior to their assembly.

It would seem that the firm first tackled the problem of barrel making before going on to the frame casting, & etc. A stock pile of partially finished barrels were made up awaiting their wedding to the brass frames which had not then been cast.

It is my guess that the stamping of these barrels with the firm name was casual in the extreme. After the stock pile of 2,650 barrels had been completed, some employee was given the die with the name "Spiller & Burr" (tail of the final 'r' broken), and told to stamp them. As a consequence, those barrels within easy reach were stamped, while those not so near to hand were untouched. The accessable ones stamped, were the ones first used when the assembled revolvers were forthcoming. Later when the firm was taken over by the C. S. Govt., a number of barrels were found to bear the firm's name, but were used indiscriminately.

## THE C. S. STAMPING

As regards the "CS" stamping found on most if not all, this was evidently applied by the C. S. Ordnance Inspector after the pistols had been received and "proved." This stamp showed they had been inspected, had passed "proof," and were official Confederate property. Oddly enough at no point in the Macon Armory records is there any clue as to when or why this stamp was applied.

The nearest approach that can be found is in Vol. #35, which deals with the Record of Repairs, and Machinework performed at the Macon Armory, Oct. '64 to March 1865. In it is the single entry that one I. H. Otto is credited with time, "stamps C. S. A. Repairs of arms."

Of those revolvers which I have been fortunate to examine, most have carried a "cryptic mark" in the form of one or two letters stamped on the frame under the grips, and also the same letter or letters stamped on the cylinder between the nipples. On the pistols examined this stamp has been "G," or "GG," and like the Griswold & Gunnison, one can only surmise that this must apply to the person who assembled the arm. Once again the records give no clue as to this stamping or to its meaning. A very careful examination of the names of the workers, however, reveal no one whose initials are "G. G.," or whose name starts with "G" was connected in any way with the assembly, or reassembly of revolvers. For the time being at least, this must remain a mystery. Vol. #39 will only add to this mystery. This volume contains a listing of all machinery and tools at the Macon Armory, listed in full and complete detail—even down to lead pencils, of which there were two. In the listing of "Tools Pistol Dept." is found the following, "stamp letters, 17, valued at $229.50." One can not help but wonder why there were only 17? Why not 26 for the whole alphabet, and also the value appears excessive?

Vol. #64, is a "Record of Tools and Machines Manufactured, Macon Armory, 1863-65." A notation contained therein only heightens the mystery. It shows; "figures or numerals, stamps for the Pistol Factory - 13, value $26.00," and "Stamps for pistol - 2, value $8.50." Both notations are dated March 1863, at which time the Pistol Factory was in Atlanta, and being operated by Spiller & Burr.

The writer has no idea as to what should constitute the "figures or numerals, stamps - 13," unless they consisted of the numerals 0 through 9, the Spiller & Burr Die, and the letters C. & S. This would account for 13 stamps, but what of the latter referred to "stamps for pistol - 2"?

In connection with the numbers of serials stamped on these guns, a peculiarity is noted in that the serial is always stamped twice on the frame—under the trigger guard plate, and again on the frame butt. The reason for this duplication is not apparent, as it would seem that either one or the other would serve the required purpose.

## COST OF REVOLVERS

Considering that on today's market a Spiller & Burr revolver approaches $400.00, and apparently rising steadily, it might be interesting to observe the cost of Sept. 7, 1864 when Burton wrote Gorgas,

................................................"Cost of revolving pistol as manufactured at this Armory at present;
Cost of materials—$19.59
Cost of labor & supervision—$34.62
Interest on capitol investment
$200,000 at 8 per cent per annum—$8.00
Total cost of pistol—$62.21
The charge for interest on capitol investment would be much reduced if the establishment were operating to the fullest capacity of production" (vol. 29)

By the end of Sept. we find that the cost per pistol was up considerably. This can be found in Vol. #40, which deals with "Machinery & Building Accounts Macon Armory, 1862-1864," for under that section entitled "Manufacture of Pistols," we find these expenses,

| "Quarter ending 31 March 1864 | to employees—$3,389.00 |
| | to stock 1,322.32 |
| Quarter ending 30 June 1864 | to employees $19,560.56 |
| | to stock 6,393.52 |
| Quarter ending 30 Sept. 1864 | to employees $5,002.18 |
| | to stock 3,169.37 |

This makes a total of $27,451.74 to labor, and $10,885.21 for material, or a grand total of $38,839.95. As we have already established that only 592 revolvers were manufactured up to Oct. 1st, 1864, some reader who is mathematically inclined can figure out just how much these revolvers cost the Confederate States Government.

A further and more complete breakdown as to the cost can be found in Vol. #55, this being the "Price Book, Pistol Dept., Macon Armory."

"*BARRELS*—forged—$.50, annealed—$.02, 1st drilled—$.50, 1st bored—$.25, 2nd bored—$.25, 1st threaded—$.22, 1st milled—$.10, rifled—$.55, 2nd milled & tapped—$.03, 3rd milled—$.05, studs fitted—$.07, polished—$.15, blued—$.08.
*CYLINDERS* forged—$.25, cut off—$.10, faced—$.03, annealed—$.85, 1st drilled—$.40, turned—$.55, 2nd drilled—$.05, tapped—$.35, nicked—$.06, polished—$.06, blued—$.04.
*LOCK FRAMES* cast—$.35, drifted—$.20, 1st drilled & tapped—$.30, faced—

$.18, 1st milled—$.25, 1st filed—$.30, 2nd milled—$.15, 3rd milled—$.20, 4th milled—$.15, 2nd drilled & tapped—$.15, interiors fitted—$.80, 2nd filed—$2.80, 3rd drilled—$.04, barrels fitted—$.10, stocked—$.80, polished—$.80.
Pistols—assembled $1.00
Pistols—proved .50
Pistols—reassembled .25."

## MEASUREMENTS

As far as can be determined, there is no essential difference between the revolvers made at Atlanta, and those made at Macon. In appearance they are the same except that possibly the later ones contain a greater angle between the grip and the frame. The general measurements of all appear the same.

*BARREL*—Octagon. Some stamped "Spiller & Burr" reading from barrel to cylinder. Some are unmarked except for the serial which is under the loading lever. Barrel length 6⅛ inches. End of barrel rounded.

*RIFLING*—7 lands and grooves, right gain twist.

*CALIBRE*—muzzle .3675, breech .3680, cylinder .3700

*CYLINDER*—made from twisted iron, lateral "faults" plainly visible. Drilled for 6 shots. Cones have the English thread rather than the standard Colt thread.

*LENGTH OF CYLINDER*—1-25/32 inches.

*DIAMETER OF CYLINDER*—1½ inches.

*SAFETY DEVICE*—Hammer engages a slot between the cones on the cylinder.

*GRIPS*—Two piece walnut. Some are fitted with exceptionally fine burled walnut grips, such as the "sample pistol" and #983. Brass washer on either side to hold screw.

*FRAME*—Of cast brass, varying in color from yellow to red, due to the many sources of brass and copper used. Most are stamped "C.S." on right or left side. Front width—9/16", rear width—51/64".

*LOADING LEVER*—Similar to the Whitney model. Later models have a pin set in the barrel lug to prevent the lever from moving horizontally.

*FORESIGHT*—Usually a brass cone set in barrel, although noted to be a steel pin carefully inset on a steel plate on the "sample pistol" and a german silver knife sight on #983.

*OVERALL LENGTH*—12⅜ inches.

*SERIAL NUMBERS*—Stamped on major parts, although some cylinders are found without serial numbers. The twisted iron cylinder here prevents some unethical person from buffing off the serials on a Whitney and "planting" it in a Spiller & Burr.

## ALMOST THE END

The story of the Spiller & Burr has been told, but in the last official dispatch from the "pistol factory" one can not help but find an inspiration.

To the common soldier of the Confederacy, the South was invincible, and to many, the surrender at Appomattox came as a genuine surprise. While this attitude was common to the combat trooper, the true state of affairs was not unknown to those higher up. Particularly was this so in the Ordnance Department, where realizing Richmond was bound to fall, Gorgas had removed his department from that city to Danville, Va., early in 1865.

April 7, 1865, was a sorry day in the history of the Confederacy. Richmond had been evacuated. The Confederate cabinet was on its way to Danville. The retreating Army of Northern Virginia, under General Robert E.

Lee, crossed the Appomattox River at High Bridge. Near Farmville, Va., the worn and hungry Confederates turned upon their pursuers and repulsed an attack by the Miles Division. General Smyth of the Union Army was among those killed. Gen. Fitz Lee's cavalry attacked the Union cavalry under Gen. Gregg, drove them back and captured Gen. Gregg. It was on this day that General U. S. Grant wrote to Gen. Robert Lee demanding the surrender of the Confederate Army, to which Lee replied that in his opinion, the necessity of surrender had not arisen.

On this black day, Col. James H. Burton, could not know the full extent of the tragedy that was befalling the Confederacy, but being a man of foresight and intelligence, there is no question but that he was aware of what the future held for the Confederacy. Whatever faults he may have possessed, Col. Burton had the virtue of high loyalty, and it was this last trait that must have been foremost when he sent the following wire to his chief, Col. Gorgas, in Danville on April 7, 1865;

"The Macon Armory is setting up the pistol machinery to duplicate that part lost at Columbia, S. C., and will start operations as soon as possible."

A man who could write such a wire at such a time, to a "chief" who was chief in name only, on a subject which pertained to future events for a nation that had no future, surely must have had his chin high in the air, and courage deep in his heart. Such a wire is the Southern equivalent of "Damn the torpedoes—full speed ahead."

Remember such men, fellow Americans, and be proud, deeply proud of your heritage.

THE END.

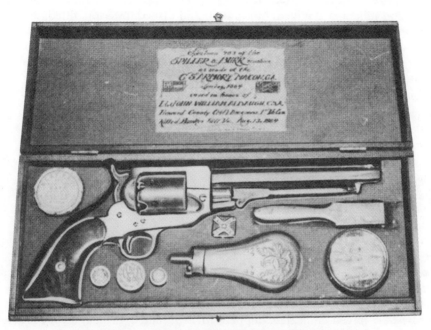

Cased Spiller & Burr #983 *(from the collection of the author)* Note similarity between the grips of this revolver and those on the "Sample" Revolver.

# APPENDIX

## *MANUFACTURE OF THE VARIOUS PARTS OF THE REVOLVERS*

Volumes #1 & 2, deal with the "Pistol Factory, Macon Armory" and give us a pretty clear insight as to the actual operation within the factory. Following is a list of all parts that went into each revolver, and the various steps taken to make each part:

"*Guards*—cast, 1st milled, second milled, drilled, 3rd milled.
*Lock Frames*—cast, drifted, 1st drilled, 1st threaded, faced, 1st milled, 1st filed, 2nd milled, 3rd milled, 4th milled, 2nd drilled, interiors fitted, 2nd filed, 3rd drilled, barrels fitted, stocked, polished.
*Barrels*—forged, annealed, 1st drilled, 1st bored, 2nd bored, 1st threaded, 1st milled, rifled, fitted to frame, 2nd milled, 2nd drilled, 2nd threaded, 3rd milled, polished, blued.
*Hammers*—forged, annealed, burred, 1st milled, drilled, 2nd milled, filed, taped, polished, hardened.
*Main Spring*—forged, annealed, filed, bent & set, tempered, polished.
*Double Spring*—punched, split, filed, set, tempered.
*Trigger*—forged, annealed, 1st milled, drilled, 2nd milled, filed, hardened, polished, blued.
*Sights*—milled and cut.
*Catch Levers*—forged, annealed, 1st milled, drilled, 2nd milled, filed, 3rd milled, tempered, polished.
*Trip Fingers*—forged, annealed, 1st milled, drilled, 2nd milled, filed, tempered, polished.
*Loading Levers*—forged annealed, 1st milled, 2nd milled, filed, 1st drilled, 3rd milled, joints fitted, polished, fitted hardened.
*Lever Catches*—forged, 1st milled, 2nd milled, filed, polished, hardened.
*Rammers*—forged, annealed, cupped, turned, milled, drilled, filed, hardened, polished.
*Lever catch studs*—forged, annealed, turned, filed, hardened, polished, blued.
*Cylinders*—forged, cut off, 1st faced, 1st drilled, 2nd faced, turned, 2nd drilled, tapped, nicked, polished, hardened, proved.
*Thumb bolts*—forged, annealed, 1st milled, 2nd milled, filed, drilled, tapped, hardened, blued.
*Double Spring Screws*—milled, cut, set, hardened, polished, blued.
*Hammer screws*—forged, annealed, milled, cut & set, hardened, polished, blued.
*Stock screws*—milled, cut & set, hardened, blued.
*Guard screws*—milled, cut & set, hardened, polished, blued.
*Trigger screws*—milled, cut & set, hardened, polished, blued.
*Thumb bolt screws*—milled, cut & set, hardened, polished, blued.
*Trip finger screws*—milled, cut & set, hardened, polished, blued.
*Loading lever screws*—milled, cut & set, hardened, polished, blued.
*Hammer rollers*—milled & drilled, filed, hardened.
*Lock screw nuts*—cast, milled & drilled, tapped, punched.
*Lock screw washers*—cast, milled & drilled, tapped, punched.
*Links*—punched, drilled, filed, hardened.
*Cones*—forged, 1st milled, drilled, 2nd milled, tapped, hardened, blued.
*Stocks (pair)*—sawed, filed & fitted.

*Center pins*—forged, annealed, turned, milled, drilled, tapped, filed, fitted, polished, hardened.
*Lever catch spring*—bent & set.
*Trip finger springs*—bent & set.
*Pistols assembled*—proved, and reassembled."

Vol. #112, is entitled "Account & Time Book, Pistol Factory, Jan. 1863 to Dec. 1863." In it are listed the various workmen at the time the plant was operated by Spiller & Burr in Atlanta. Believing that a portion would be of interest to the reader, the following is extracted therefrom.

### SOME OF THE JOBS, & THE WORKMEN OF SPILLER & BURR, Oct., Nov., Dec. 1863

Fitting barrels to frames—Thomas Burton
Milling frames—John Berry
Drilling cones—A. D. Ruede
Assembling & testing pistols—A. J. Bass
Filing hammers—John Breen—filing component tools for shop.
Night watchman—Michael Regan
Milling triggers & levers—Anthony Bice
Case hardening—Denis Sulivan
Casting frames—Blankenship
Assisting in polishing—G. D. West
Repairing tools—R. G. Scott
Fitting frames, links for loading lever, asst to testing pistols—G. Badger
Clean frames, making cylinders—Lefountain
Repairing tools—Thomas Smith
Repairing tools forge hammer—W. E. Burns
Polishing frames, repair tools—I. A. Tuttle
Drilling hammers—H. Bradford
Polishing frames—Howard Thomas
Testing pistols, asst. inspector—James S. Claspy
Milling triggers & components—I. P. Samuel
Turning cylinders—A. Calder
Repairing tools—Dwight Wing
Night watchman—W. Clably
Milling frames and day work on catch levers—Henery Wright
Milling 'L' catches, filing frames, rifling barrels—H. Downs
Stocking pistols—Pesse Willburn
Milling hammers & loading levers—William Eaton
Assembling pistols—E. Campbell
Repairing tools, turning cylinders, Engineer—Allen Friss ·
Drilling cylinders—F. S. Gregory
Milling hammers, center pins and loading levers—John T. Hutchinson
Assembling pistols, filing frames, fitting barrels in frames—Bartholemew
Milling components—Edw. Baldwin
Drilling cylinder—L. Bice
Forging loading levers, thumb screws, drawing steel for cones, work with Capt. Spiller—T. M. Horton
Cutting cylinders to length—James Sloman, testing & inspecting pistols
Filing thumb catches, fitting frames, filing hammers—T. E. Hall

Milling components—William Robinson
Filing frames and hammers—Wade Hall
Testing pistols—Martine Blankenship, milling frames & LL catches
Countersinking cylinders, making cylinders, taping, turning & facing cylinders—David Jones
Filing hammers—A. H. Mandeville
Stocking pistols, day work for Capt. Spiller—I. T. Jones
Filing hammers—A. M. Lamb
Assistant in inspecting room—Thomas Jones
Filing hammers—William Bowles
Filing triggers & loading levers, trip fingers, filing center pins—C. A. Johnson
Reaming cylinders, assisting in foundry—John Scantell
Burring hammers, drilling l. levers, assisting in testing, end reaming on barrels—John Kenedy
Countersinking barrels, cutting barrels to length—James Lawske
Drilling cylinders—John Lefountain
Case Hardening—Henry Jones
Stocking pistols, day work for Capt. Spiller—George Fisher
Smith's helper—John Savage
Day work in shop (making springs, milling frames & guards, polishing)—John Lynch
Assisting to test pistols, cutting off cylinders—Henry Proctor
Assembling pistols—T. I. Tair
Milling components—Henry Swope
Assistant in inspecting—John Zacknew
Drilling cylinders—Lewis Lanier
Casting frames—I. Q. Morris
Filing frames—W. O. Rankins
Milling frames, cutting off cylinders for shop—W. C. Moore
Trip finger fitting frames, sear spring, filing main spring—I. E. Rankins
Filing catch levers, loading levers, hammers—Miss T. A. Kelley
Drilling cylinders—I. H. Plunkett
Cleaning frames—Daniel Mohaly
Inspector—I. B. Myers
Testing & inspecting pistols, milling frames—James Mahool
Foundry work, fitting frames—W. McElmore
Filing frames—John Midleton, day work on engine
Filing triggers, Engineer—John McGee
Polishing components—M. N. Nicholson
Milling components—Charles Owens
Boy assisting in foundry—Thomas Ivory
Fitting hammers in frames—James Reeves
Filing Frames—I. N. Butler
Milling frames & guards—A. Richardson (making cylinder work for Lafountain)
Drilling frames—Robert Alley
Foreman—O. H. Phelps
Forging—Jas. Armstead
Assisting in foundry—James Berry.

It may also be of interest to learn from where these workers came. This information is contained in Vol. 112.

## NAMES & BIRTHPLACE OF WORKMEN OF SPILLER & BURR

M. A. Alleyn, New Orleans, La.; James McAlpin, Chatham Co., Ga.; R. J. Donnavan, Cork Co., Ireland; S. S. Branney, Stratford, Northampton Co., England; M. Hambrick, Jones Co., Ga.; Aaron Stevens, Westbrook, Middlesex Co., Ct.; G. W. Waston, Porquamins Co., N. C.; Robert Saunders, Caroline Co., Va.; J. S. Peck, Danbury, Conn.; John Burke, Waterford Co., Ireland; John Faley, Limerick Co., Ireland; William A. Wiley, Houston Co., Ga.; J. H. H. Porter, Jones Co., Ga.; Carnelius O'Connell, Limerick Co., Ireland; F. N. Horton, Cincinnati, Hamilton Co., Ohio; John Crass, Decatur Co., Ga.; William Allen, Calhoun Co., Oxford, Ala.; John Cole, Quincy Adams Co., Ill.; William O. Moore, Houston Co., Ga.; Jos. Anthony, Cromwell, England; Jill Richardson, Macon, Bibb Co., Ga.; Edwin Anthony, Cromwell, England; J. J. Wood, Exxes Co., N. Y.; McConnell, Charlotte, Va.; P. S. Rogers, Belfast, Ireland; Miss Kelley, Lauderdale Co., Tenn.; Doby, Lincoln Co., Tenn.; Strickland, Buck Co., Penna.; G. R. Badger, DeKalb Co., Ga.; D. A. Jones, Monroe Co., Ga.; Jerome Hall, Lexington Dist., S .C.; John L. Morris, Monroe Co., Ga.; John E. Rankins, Lexington Dist., S. C.; William A. Rankins, Lexington Dist., S. C.; William A. Moore, Greensboro, Green Co., Ga.; J. F. Sloman, Dover, Stafford Co., N. H.; A. D. Ruede, Salem, Stokes Co., N. C.; Henry Ruede, Marietta, Cobb Co., Ga.; J. H. Wright, Brownsville, Heywood Co., Tenn.; Edward Campbell, Greenville Dist., S. C.; Foster Richardson, Pickens Dist., S. C.; John F. Kemp, Viana, Dooley Co., Ga.; John F. Cones, Morgan Co., Ga.; William Wood, Macon, Bibb Co., Ga.; A. J. Smith, Macon, Bibb Co., Ga.; F. J. Harrington, Harpers Ferry, Jefferson Co., Va.; William J. Taylor, Columbus, Ga.; Henry Taylor, Montgomery Co., Ala.; J. J. Gay, Isle of Wight, Va.; H. W. Barr, Putnam Co., Ga.; Alex. Calder, Glasgow, Scotland; George Haszenger, Bavaria, Germany; A. G. Herrington, Harpers Ferry, Va.; Wade Hall, Lexington Dist., S. C.; Irvin Hall, Lexington Dist., S. C.; William R. Dawson "does not know where he was born"; A. J. Bass, Mercer Co., Kentucky; H. H. Herrington, Harpers Ferry, Va.; Joseph Brown, Macon, Ga.; and John M. Lynch, Mobile, Ala.

## OPERATIONS AT MACON ARMORY

Included in this section are several letters from Burton to Gorgas. They are placed here because it is believed they will be of interest only to the serious collector.

Taken as a group, they give a rather complete picture of the operations of the Macon Armory.

The first of these letters dated 8-23-62, is to be found in Vol. #20, and covers Burton's activities pretty thoroughly from the time he located in Macon (June 1862) until the date of the letter.

"Colonel:

In compliance with your instructions of the 14th inst., I have the honor to present for your information the following statement with reference to the progress of the work undertaken at this armory.

On the 1st of July ult. I made an agreement in writing with the representatives of the city of Macon, granting for the use of the C. S. Govt., the premises known as the "Old Depot" of the Macon & Western R. R., for such period as the Govt. may require them for temporary purposes at the rent of $1000 per year. These premises include about 3 acres of land, graded perfectly level—one brick bldg., 130 X 33 feet, one story high, formerly used as offices, and one wood bldg., about 150 X 20 feet, formerly used as car shed, also the use of another brick bldg., about 100 X 33 feet one story high, formerly used as an Engine & car house, and distant from the beforementioned premises about one fourth of a mile. At the time of my making the arrangement for the use of these premises the old depot bldg., was occupied by Capt. R. M. Cuyler, Comdg., Macon Arsenal, as a magazine for powder and ammunition, and it was agreed between that officer and myself that he should have the use of the outer brick bldg. for his purposes in order that I might have the old depot bldg., for the purpose of erecting therein the set of stock machinery from the Richmond armory. In order to carry this arrangement into effect it was necessary to make considerable repairs to the outer bldg., requiring nearly three weeks to accomplish. This delay retarded my progress this length of time much to my regret, but it was unavoidable. Having obtained possession of the depot bldg., it became necessary to lower the entire floor two feet in order to obtain the requisite height inside, and also to replace the wide doors in the walls of the bldg. with suitable frames & windows. Also to erect a row of posts along the center of the length of the bldg., to carry the line of shafting all of which has been accomplished, and an excellent shop for the purpose is the result. The stock machinery has all been unboxed and so far as yet as contained, but one machine has been injured in transportation here and that one has been repaired. All these machines are now in place, and much of the driving shafting, etc., erected, inclusive of the mainline shafting—and all will be ready to put into operation by the last of next week, or by the time I receive some gunstocks—none of which have yet reached here. The steam engine received from Knoxville, Tenn., along with Maxwell's machinery, has been put in complete order, and is now in place complete, ready for use. I have erected a new boiler from Vicksburg, Miss., selected from Reading's lot of machinery and have sunk a well 30 feet in depth from which I shall obtain a good supply of water for steam purposes. A brick boiler and engine house is now being erected to cover the above and will be complete in three days' time, I have also erected a brick chimney 61 feet high adjoining the boiler house. I shall get steam up as soon as the boiler house is completed—say in 3 or 4 days from now, and will then be ready to go on regularly with the production of stocks in the machined state. The failure to receive the assistance of several of the workmen from Richmond accustomed to the use of the stock machines may retard my progress to some extent but I hope to soon educate others who will replace them. It is my purpose to commence as soon as possible, the manufacture of the machinery for the new Armory, and with this view I contemplate the erection of a frame bldg., 200 X 35 feet to be used as a forging shop and will contain 12 forges. A third bldg. will also be required 90 X 33 feet, one story high, to be used as a store house for gunstocks and other stores and materials not in current service. The timber for all these bldgs. is now on the ground here; the large bldg. for machine shop is nearly all framed and the brick foundation is laid ready for erection of frame which will commence in one week's time from now. I hope to have this bldg. ready to receive machinery in about 3 or 4 weeks' time from now. The other two bldgs.—being much smaller, and but one story high—will not require long to frame, and erect, and will be ready I expect about the same time as the large bldg. The whole of the premises have been enclosed with a strong vertical board fence 6 feet high. I have employed to a great extent, negro carpenters in the framing and erecting of the bldgs. and I am pleased to state that I have found their employment very satisfactory indeed.

The machinery assigned to my use from Holly Springs, Miss., Richmond, Va., and Raleigh, N. C., has all arrived and a force is now unboxing it and cleaning and putting it in order. That from Holly Springs is in very bad condition owing to bad packing and rough usage in transportation, an inventory is being taken of everything received from Holly Springs on the completion of which I shall receipt for the property to Maj. W. L. Brown. I see no difficulty to be apprehended in the construction of the machinery for the new Armory, provided the mechanics can be obtained, many of whom are now in the army, and it will be necessary to resort to some ready means of selecting them and detailing them for special duty as mechanics in this armory. The present routine in force through the War Dept. at Richmond seems to be too slow and uncertain. Feeling the necessity of erecting the stock machiners in the shortest time possible, I directed all my energies and attention to that end, and now that this is about being accomplished, I shall at once take up the subject of the erection of the main works for the armory proper. I have requested the city to survey the site for the armory, which will be done as soon as an engineer can be found to do it on the completion of which, a plan will be forwarded to you. Also plans and elevations of the proposed buildings as soon as their character, etc., has been decided upon.

It gives me much pleasure to state that the utmost good feelings is manifested towards the enterprise. I am charged with by the citizens of this city generally, and my experience as far has been satisfactory, and I have every reason to believe that the selection of this point as a location for the armory will be attended with results satisfactory to the Govt.

I will report to you monthly the progress made at this Armory as instructed by you."

## PHYSICAL ASPECTS OF THE ARMORY

Letter #2, also from Vol. #20, dated Oct. 16, 1862, gives a complete picture of the physical aspects of the Macon Armory.

"Colonel:

I have the honor to transmit enclosed herewith, a copy of the Title Deed, conveying to the C. S., the ground donated by the City of Macon, for the purposes of an armory together with tracings of the map and profiles of the same as made and determined by the engineer employed by me to make the survey. Upon referring to the map, you will find the proposed site for the armory bldgs. indicated approximately by the outline in red ink, this being the most eligible spot for them—the ground being generally more or less "rolling" in character. The pile of bldgs. will cover an area of about 5 acres and on the spot indicated, I shall be able to secure the desired quantity of sufficiently level ground.

The branch represented as running parallel with Hazel St., is a never failing stream and from this branch I propose to derive the necessary supply of water for steam and other purposes. You will observe that I have so arranged it as to control this branch exclusively, as it is *included* in the ceded ground the whole length of that side of the tract, thus preventing any interference with the water supply by outside parties. The highest portion of the ground lies along the line of Calhoun St., and here will be the best site for Officers' Quarters. In this connection I respectfully suggest the desirableness of acquiring the Square numbered #49 on the map, which can be accomplished by purchase from the city, at a cost of $7000 to $8000. The Square contains 4 acres, and embraces the highest point of ground in the immediate vicinity, and is bounded on all sides by wide streets. This square would afford beautiful sites for quarters and were it purchased at the maximum price above quoted, the Govt. would be possessed of over 46 acres of valuable land for the purposes of the Armory at the very moderate cost of $8000, and which at present prices is worth at least 5 times that sum. I respectfully recommend that purchase of square #49 for the reasons stated and solicit your instructions at your earliest convenience in reference thereto. I am preparing to enclose the whole tract of 42¼ acres with a temporary but substantial board fence 8 feet high, and a portion of the lumber for that purpose is now on the ground—the balance is under contract and will be delivered in a week or two. As it will be necessary that some responsible person should reside upon the spot and take care of the Govt. property thereon, I propose to erect one or two frame cottages of 4 rooms each, at a cost not to exceed $1200 to $1500 each which may be rented to, say the foreman of labourers and the foreman of carpenters, at a moderate rent. There are at present no bldgs. of any description on the land herein referred to. At the present time I am considering the subject of the character and style of armory bldgs. to be erected. The difficulties of the present times will make it necessary to confine the style to one of a comparatively plain character. The supply of mechanics of the necessary high degree of skill, is so small that were bldgs. of an elaborate style attempted, I fear their erection would be entirely too slow an operation. I state this now in order to prepare you for the plans and drawings of the proposed bldgs. which I shall submit to you as soon as they can be completed, and which will exhibit a plain but bold style of architecture.

I have made a contract for three millions of bricks at the price of $11.00 per 1000 delivered at the Armory, and expect to have nearly one-third delivered this fall; the deliveries being already commenced.

The principal temporary bldg., erected upon the ground *rented* from the city of Macon is about completed and the lines of shafting are now being erected in it—and the removal of machinery into it is in progress. In a week or ten days' time I hope to have some of the machines running. The framing of the two minor bldgs. (Smith's shop and storehouse) is in progress, and will not require long to complete being but one story high. I have received what portion of Reading's Machinery I desired but I shall be much inconvenienced in consequence of numerous and important parts of some of the machines being lost. The machines are useless until these parts can be replaced. I have also received some machinery from Knoxville, Tenn. (Maxwell's), most of which is in the same condition as Reading's. I have received but a small portion of the machines, etc., etc., enumerated in the list I received from you before I left Richmond, being that which was turned over to the Govt., from Knoxville.

Capt. Wright states that he has, however, forwarded all the machinery he received from that source.

The mfgr. of gun stocks at the armory progresses currently, and satisfactorily and in order to secure a supply of the Rich. Armory as early as possible, I am running that dept. extra hours, but morning & evening. I shall soon forward 1000 stocks in charge of special messenger.

I experience much difficulty in obtaining the mechanics I desire (smiths & machinists) and the effect of this difficulty upon my future operations I shall make the subject of a future letter to be addressed to you at an early day, and in which I shall take occasion to offer some suggestions by which the difficulty may be possibly overcome."

The next letter dated Nov. 15, 1862 (also from Vol. #20) deals almost entirely with the problem of gun stocks, and is written by Burton in answer to a request from Gorgas to manufacture gun stocks for the model 1842.

"Colonel:
I have the honor to acknowledge the receipt of your telegram of the 14th instn. informing me that 'all the gunstocks received will be exhausted before the end of the present month.'

In reply I beg to state that every possible effort is being made to increase the production of stocks to the utmost extent. That dept. is now working from sunrise until 10 o'clock P. M., inclusive of Sundays, and if it be possible to accomplish it I will arrange to work all night. Most of the workmen employed on this machinery are new and inexperienced, but will become more expert as they become accustomed to the work and will be enabled to turn out more stocks than heretofore.

I now have about 700 stocks ready for packing, and as soon as the number is increased to 1000 I will forward them to Richmond in charge of a special messenger. It is fatal to attempt to forward them in any other way, and I have thought it not worth while to forward a smaller number than 1000 at a time. If you think otherwise please so instruct me. I think I shall be able to deliver 1000 stocks in Richmond by about 26th inst.

I beg to acknowledge the receipt this morning of your letter of the 11th inst., in which I am instructed 'not to hazard the reduction of stocks model 1855 below 1000 per month and the 2000 now needed additional.' Since writing my letter to you of the 7th inst., on this subject I have investigated it more closely and taking all the circumstances into consideration I have concluded it would not be politic to attempt at present the alteration of the machinery so as to fit it to the production of stocks of the model 1842.

The machine cannot be altered without suspending their operation more or less, which of course would decrease the production of stocks of the model 1855 to an extent very undesirable under the present pressure. In addition to this, I have as yet been able to procure the services of but few machinists of sufficient skill to employ upon work of this class. Looking at the subject in all its bearings, I cannot think it advisable to attempt the alteration at present and therefore respectfully advise that it be postponed until a more favorable opportunity presents itself. In the meantime I will push on with the stocks model 1855 and will be pleased to receive your further instructions. I may perhaps be able to complete one of the machines for rough turning stocks, re'cd. from Holly Springs, and if successful will then be able to at least expedite the object in view to this extent. I will make the effort at once, and advise you of the result as soon as possible."

Jan. 10, 1863, Burton again writes Gorgas a lengthy letter, giving full details as to the progress being made at the Macon Armory. Note Burton's subtle request to be sent to England, which request was acted favorably upon by his chief. (vol. #20). The "Razeed Carbine" to which Burton makes so much mention, is the Richmond high wall carbine made from cut down or "razeed" stocks and barrels.

"Colonel:
I have the honor to present the following statement of work done at this Armory since my last report of Oct. 16th, 1862.

The temporary bldgs., erected on the ground rented from the city of Macon, has been quite completed, and all the machinery has been repaired, put in working order, and erected, ready for use. The machine-shop, 200 X 25 feet, 2 stories high, is now full of excellent machinery, propelled by 400 ft. of main line shafting & pullies. This shop is capable of employing 150 machinists. The smith's shop has also been completed, 100 X 35 ft., 1 story high, fan blast, and blast pipe of wood laid down, and 8 cast iron forges of a new pattern gotten up at this Armory, erected, and ready for use. A store house for stocks, and other stores & materials, has been erected & is now occupied. All of these bldgs. being of frame have been well whitewashed, & present a neat appearance. The stock machinery is kept steadily at work early & late, and regular monthly supplies of stocks have been, and will continue to be forwarded on to Richmond. A messenger will leave on Monday evening next, in charge of 1008 stocks, for rifled musket, and 432 for Razeed carbine for the Richmond Armory. The fabrication of machinery has been commended, and will be pushed forward as rapidly as the limited force at my command will permit. In pursuance of your authority as contained in your letter of the 26th Oct., I have purchased block #49 plan of the city of Macon, containing 4 acres at the price of $8010, and have rec'd. title deed for the same, a copy of which I enclose herewith for your information. It will afford an excellent site on which to erect officers' quarters. In accordance with the conditions of the grant of land for Armory purposes by the city of Macon, I have constructed a good road along the line of the Macon & West. Rwy., in order to divert the travel of vehicles which previously was immediately across the armory grounds. This enables me to enclose the grounds which is now in progress, the posts being nearly all planted for the temporary fence. The whole fence will be completed in a few weeks. Permanent corner stones of granite properly marked with sunk letters have been planted at the angles of the armory grounds. The work of grading the ground for the site of the armory bldgs. has been commenced but a sufficiently large force cannot be employed for the want of tools (shovels & picks) which I have not yet been able to procure. A contract has been entered into for the stone work, of the foundations of the armory bldgs., and the blasting of the necessary stone has been commenced. A contract has also been made for the supply of 5000 barrels best quality lime at a reasonable price to be delivered as required. The framing of two cottages to be erected on the armory grounds has been commenced and is well advanced. Preparations are in progress for the introduction into the Armory grounds of a siding from the M. & W. Rwy., which will greatly facilitate the delivery of materials for bldg., etc. At the present time, I am receiving proposals for the supply of bricks but I find parties unwilling to contract for articles to be delivered next summer, except on such terms as amount almost to extortion. If I find that I can not make satisfactory contracts for bricks I must again make an effort to arrange to make them myself. I much prefer to have them supplied by contract, however, for several reasons. A conditional contract has been entered into for the supply of roofing slate, the condition being that the Govt. will detail four men from the army whose services are necessary to the contractor.

At the present time I find it impossible to get supplies of coal from Chattanooga in consequence of the difficulties of transportation. I have orders there now for 2500 bushels of coal, none of which I can get, and I am compelled to borrow from Major Cuyler, Comdg. Macon Arsenal, who is also running short of coal. I shall be glad if this difficulty can be removed so that supplies of coal can be obtained, at least sufficient to keep us going.

In my report to you of the 16th Oct., last, I referred to the great difficulty experienced in the effort to obtain the services of competent machinists, and blacksmiths, in sufficient numbers. I regret to have to state that the same difficulty continues to exist to the great detriment of the progress of the work I am charged with in part viz: the construction of the necessary machinery for this armory. I have feared this difficulty from the beginning as being one entirely out of my power to overcome. I could now employ 170 machinists, and smiths. I have but about 35, and cannot obtain more and the Sec. of War, declines to grant details of men from the Army when confronting the enemy, under which circumstances, the greater portion of our Army is now placed. If I could command the workmen, the machinery can be constructed here. If the workmen can not be had the machinery can not be made here. The question which suggests itself then is, how can it be obtained? I have given this subject much thought and the result is my conviction that it can be obtained from England. I know of large machine shops in England in which I had much of the machinery for the Enfield Armory constructed, and I have no doubt but that contracts could be entered into with the proprietors of these shops for the construction of much of the machinery required for this armory. I regard it as absolutely necessary to go to England for the large steam engines required for propelling the armory machinery as there are no shops now in the Confederacy capable of constructing such as are required, except the Tredegar Iron Works at Richmond, which are necessarily otherwise employed. The machinery for barrel rolling and welding would also have to be obtained fom England as it has never been made in this country. Looking at the question in all its bearing, I am of the opinion that the best course to pursue will be to send out to Europe some thoroughly competent person with power to enter into contracts for the machinery required and also to obtain glass, hardware and other materials required for building purposes, not now to be had in the Confederacy. In the meantime, the machine shop of this armory can be kept employed in the construction of such machines as can be made here such as tilt hammers, power punches, etc. the machinery constructed abroad can be run thru the blockade with as much facility as other goods and with as little risk of loss, or it can remain in Europe until peace is declared when it will be at once available. In this way the construction of the machinery abroad would be contemporaneous with the erection of the bldgs. here, which will require at least one year to erect under the most favorable circumstances. I may add that I have consulted with several prominent officers of the Ordnance Dept., on the necessary of adopting this policy, among whom I name Col. Rains, Gen. Huger, & Major Cuyler, and they all agree with me in the views herein expressed.

Respectfully requesting your earnest consideration of these recommendations."

Oct. 30, 1863, being a report of Burton's visit to England. (vol. #31)

"Colonel:
In submitting this, my final report on the subject of my late mission to England, I have deemed it proper to furnish you with copies of my entire correspondence in relation thereto. I have the honor therefore, to enclose herewith for your information, copies of the correspondence referred to, omitting only a copy of the contract between Fraser, Trenholm & Co., of Liverpool, and Greenwood & Battey (Batley (?)) of Leeds England; the receipt of which from England, you have acknowledged together with my three official letters reporting

progress at various periods during my absence.

My last letter from England, was dated Aug. 7, 1863, at which time satisfactory progress had been made with the machinery contracted for by Greenwood & Batley and the same favorable report of the progress made up to the time of my departure from England (5th Sept.)—sic. may be made, and I have no doubt but that the condition of the contract as regards periods of delivery will be strictly complied with. As I could not remain in England to inspect the machinery as required by the conditions of the contract, it became necessary to appoint some competent and reliable person to perform that duty. I accordingly with the consent of Major Caleb Huse, made an arrangement with Mr. James Davidson—mechanical manager of the Royal Laboratory, Woolwich Arsenal, who I have known for a number of years as being a thoroughly competent machinist—by which the machinery will be inspected by him, in lots as manufactured and ready for inspection. The compensation to be allowed him for performing this duty is $250. sterling, for the entire lot of machinery contracted for with Greenwood & Battey.

I brought over with me all the plans and drawings necessary to the putting down of the foundations for engines, geering etc., and for the barrel rolling machinery; all of which can now be proceeded with as fast as local circumstances will permit. Before leaving England, plans, and drawings of most of the special and most important machines had been prepared and were approved by me; and I left with the contractors all necessary instructions to enable them to proceed understandingly with the work, and I know of no reason why the contract should not come to the most satisfactory conclusion.

Some weeks prior to my leaving England, Major Huse, requested me to procure information concerning the construction of machinery required for laboratory purposes and for the manufacture of gun carriages. This request I so far complied with as to obtain lists of the most useful machines and apparatus now in use in the Royal Arsenal, Woolwich, and tenders from the mfgrs for the construction and supply of said machines; but up to the time of my departure from England, Major Huse was unable to provide the funds, and consequently the contracts were not entered into. At the request of Major Huse, I left all the papers and information in relation to this machinery with him, and it was his intention to contract for it as soon as funds were available for this purpose.

I also obtained a list of machines comprising a full sett (sic) ofor (sic) the mfgr of powder barrels such as has been for some time in use in the Royal Arsenal, Woolwich, and a tender for the construction and supply of such machines. The cost of the full sett amonts to 1260 pds, sterling only, and I am quite of the opinion that such a set of machines should be associated with the powder mills at Augusta, Ga.

I saw samples of the barrels made by this machinery and they were certainly superior to any handmade barrels I ever saw, besides being made at very much less cost. I therefore respectfully recommend the purchase of the sett of machinery for this purpose. Enclosed herwith I enclose copies of the tenders for the supply of the machinery selected by me for the mfgr of gun carriages, and for laboratory purposes, and also for the plant of machinery for making powder barrels, all of which I trust you will find satisfactory. The prices I regard as very reasonable.

I regret to have to report that I was unable from want of funds to purchase any of the bldg materials required for the purposes of this Armory or for any other Ordnance establishment. Major Huse could not furnish me with even the small sum of 2800 pds with which to purchase the window glass, sheet copper, etc. required for the completion of the new bldgs in process of erection at this Armory. A reference to the letter of Major Huse, dated Aug. 19th, will explain his inability to furnish any funds for this or any other purpose. At his request, I left with him a list of the materials required for this armory, and it was his intention to purchase and forward them as soon as he had the necessary funds. I arranged with Messers Fraser Trenholm & Co., for the shipment to Bermuda of the machinery in lots as delivered, and as opportunities offered of shipment. It was considered politic that no *entire* cargo should consist of machinery. I informed Major N. S. Walker the agent of the Ordnance Dept. at Bermuda, on the subject of receiving and carefully storing this machinery as it arrives; but it would be well for you to give him written instructions in relation thereto, in order to insure the carefully handling and storing of the machinery, that it may be eventually arrive here in good order. The custom seems to be to handle cases very roughly and in the case of the expected machinery special precaution should be taken to prevent such rough handling, otherwise disaster will be sure to result.

Mr. Charles Lancaster, the inventor of the well known Lancaster Gun, etc., kindly presented me with plans and drawings of a new system of iron plating for ships and shore batteries which he has invented and which is now being applied to some batteries in process of erection. He also gave me a report on experiments made with a 9 pd field gun, rifle on his principal of oval bore, together with a photograph of the gun mounted. All of the above are respectfully submitted herewith, as also several printed reports of late date on the subject of guns and small arms. These last are in the form of Blue Books, printed by order of the House of Commons and they throw much light upon the subjects treated therein, and they will be found very interesting. I also purchased with my private means one copy in 3 vols. of a late edition of the Aide memoire to the Military Sciences, one copy of a recent publication entitled Col. Anderson of the mfgr of Gunpowder and one copy of a pamphlet on The nature of the action of fired Gunpowder, also two micrometers, one large and one small, for measuring objects to the one thousandth part of an inch. Should the Ordnance Dept. desire to purchase all the above, I shall be glad to dispose of them at cost price.

A short time before I left England, I visited the Govt. Powder Mills at Waltham Abbey and thru the kindness of the superintendent, Col. Askwith, R.A., I was permitted to visit all parts of the works; in fact Col. Askwith spent a half day in accompanying me over the establishment and making explanations in relation thereto. He also kindly furnished me with samples of the various woods and charcoal used in the mfgr.—specimens of the Mill Cake and Press Cake, which I respectfully submit herwith. Col. Askwith understanding that a large establishment for the mfgr of gunpowder had been erected by this Govt., expressed a desire to be furnished with a general ground plan of the works showing the arrangement of the various bldgs, and processes of mfgr. In consideration of the courtesy extended to me by Col. Askwith, I respectfully recommend that his desire be complied with, and I shall be pleased to forward to him any drawing or information you may be pleased to furnish me with for that purpose.

I also visited the Royal Small Arms Factory at Enfield, and was kindly shown over the works by the superintendent, Col. W. M. Dixon, who received me most courteously. At these works preparations *were being made for the mfgr of 8000 rifles on the Whitworth principal* of the hexagonal bore of barrel with a view to giving this principal a thoro test in the field. It is probable that some change will be made before long in the character of the rifled arms required for British service, but up to the present time no decision has been arrived at in this point. A late report (herein submitted) is decidedly in favor of the Lancasters principal of oval bore for small arms.

In consequence of the want of funds I regret that I was unable to purchase specimens of the arms adopted in the various European services, as instructed by you to do. The gunmakers of Birmingham at last acknowledge the superiority of the system of mfgr of small arms in quantity by machinery have formed a joint stock with a view to the mfgr of military rifles by that system, and the principal bldgs of the factory are already erected and much of the machinery completed.

At the time of my visit to Birmingham the military gun trade was at a standstill almost from want of orders—the U.S. Govt., was not then being a purchaser.

In concluding this report I have much pleasure in testifying to the kind reception I universally met with in England, and also to the desire manifested by mfgrs & others to serve the C.S. Govt. in their respective specialties.

My mission was to me a pleasant and satisfactory one, and the only regret I have to express is that the necessary funds could not be provided for the purchase of materials required by this, and other mfgring establishments of the Ordnance Dept.

I have the honor to be.
Colonel.
Respectfully, your obdt. servt."

Oct. 23, 1863 another full report on the progress at the Macon Armory. (vol. #31)

"Colonel:
I have the honor to present for your information the following statement of progress at this armory as I find it on resuming my command on my return from Europe.

## BUILDINGS

The erection of the New Armory bldgs has not progressed as rapidly during the past summer as I had hoped. It was my expectation that the main centre building (2 stories) would have been ready for the roof by this time, but I find that only one half of the length has been raised to the height of the 1st story ready to receive the joists and girders for the 2nd floor—the laying of which will be commenced almost immediately. In the meantime the workmen have commenced the laying of brick on the other half of the building, which will be pushed as rapidly as the limited force of bricklayers will permit. The delay is due for the most part I find to the non-delivery of bricks by the contractors in time to allow of the laying of them early in the past summer as designed. The bricks are now being delivered however more rapidly, than they can be laid and there is great difficulty in obtaining a sufficient number of bricklayers. There are now employed only 13 bricklayers whilst thrice that number could be employed to advantage. I am now making every effort to increase the number but as yet without much success.

Good progress has been made with the carpenters work for these bldgs., which is much in advance of the brickwork. All the window frames required for both stories of the large bldg. in hand are completed, ready for setting in place, much of the framing is done for floors etc., and the sash and doors will be made this fall and winter, quite in time for the other work to receive them.

The progress of the stone foundations has been very satisfactory, and the contractors for this part of the work has done well. The entire foundation walls of the main centre building have been completed in good time, and good progress has been made with the laying of the foundation walls of the rear range of buildings for smithy etc., the erection of the superstructions of which must be deferred until next year. Considerable grading and filling has been done around the buildings and at other points on the grounds.

The contractors for lumber have generally fulfilled their contracts and so far, there has been no delay resulting from this cause. The contractor for slate for roofing is much behind in his deliveries but I have reason to believe he has done the best he could with the limited number of slate quarriers he has been able to secure. The Govt., has not assisted him with detailed men as stipulated in the contract with him to comply with his engagements under it. At the present time, there is some difficulty in obtaining cars to transport the slate in consequence of the great demand for transportation of quartermasters and other stores for the supply of Gen. Bragg's Army. This difficulty must continue to exist as these supplies are required from this direction, which will probably be during the whole of the coming winter.

The plant of wood working machinery erected and put in operation during the past summer has been productive of much saving in the cost of the work done, and the work has been greatly facilitated. This machinery has been temporarily erected in the new Proof House, and is driven by a 15 HP steam engine purchased for the purpose. The whole of this arrangement and its results have been very satisfactory. During the current year the negro mechanics and laborers employed at this Armory have been in all cases hired by the day, but from present indications it is not probably that they can be hired next year. I have therefore determined to erect temporary negro quarters in order to be prepared to accommodate such negroes as I may be compelled to hire by the month year after Jan. 1st, 1864, and who will have to be provided with food and lodgings. I understand from Col. Cuyler that you have already granted authority for the employment of negroes this way.

The erection of a frame house of 8 rooms for the accommodation of the Master Machinist and his family is about being commenced. I believe by your authority.

It will be impossible to complete the main Armory Bldg., now in progress before next Spring, for the additional reason that the materials—glass, sheet copper, hardware, slating nails, etc which I expected to procure in England have not been supplied in consequence of the inability of Major Huse to furnish the necessary funds for this purpose.

## MACHINERY

The fabrication of the machinery reserved to be made at this Armory has likewise progressed slowly: partly in consequence of the small number of machinists, available, and partly in consequence of a portion of them having been employed in the mfgr of projectiles for Charleston.

During the past summer, the following machines have been completed and

partly finished, viz:
4 machines for the 1st boring barrels, completed.
6 machines for the 1st boring barrels, partly completed.
6 machines for the 2nd boring barrels, partly completed.
3 punching presses (geared), completed.
1 cast iron trough etc., for small grindstone, completed.
2 upright jig saws and frames, completed.
Iron work for 4 trip hammers, completed.
6 spindles & pulleys for large grindstones, completed.
40 pairs hangers for counter shafts, partly completed.
1 machine for making function tubes nearly done.
Patterns for furnaces for annealing and casehardening.
Forges, stock machines, pulleys etc. etc. (sic)

In consequence of the high rate of wages now paid to machinists the fabrication of machinery is very expensive indeed, so much so as to induce me to recommend its mfgr in England, as being the cheapest means of providing it. *Before my late visit to England I was not aware that the machines required for the mfgr of the gun stock were made in that country,* but I found that the mfgr of these machines had also been introduced there, and I saw some in process of fabrication for a private concern, the workmanship of which appeared excellent. I am of the opinion that the machines of this class required for this Armory can be procured from England at less cost than they can be made for here, and with your consent I will request tenders for them by the mfgrs. I am decidedly in favor of limiting operations at this Armory to the production of such tools, fixtures, etc., as from their special character cannot be obtained from abroad—the difficulty and expense of the mfgr of everything has become so great as to strongly recommend this course to me. I respectfully request your opinion and instructions on this question. If the funds derived from the sale of cotton sent out in the vessels belonging to the Ordnance Dept., were confined to the purchase of such articles and materials as are required by the Ordnance Dept., they would be ample in my opinion for all the required purchases abroad. But I was informed in England that this had not been the case, and hence the want of funds for the purchase of materials required for this Armory.

The difficulty of doing mechanical work of any kind has become so great as to make even ordinary progress in any department impossible, and at this Armory, it pains me to have to report such slow progress. But this slow progress is inseperable from the present unfortunate condition of the whole country and hence it is almost vain to strive against it. We can only do the best we can the face of the difficulties (sic) and rest content with the result.

I append hereto for your information a statement of the number of workmen of each class at present employed at this Armory, and the rates of wages paid to each. The wages seem high but if the cost of the necessities of life continues to increase—as is probable, the present rates of wages will have to be increased also. This might be avoided however to a certain at least if an arrangement could be made by which the workmen could have the privilege of purchasing of the commissary; provisions at Govt. prices. I am told that this is practised at some of the Arsenals, and I respectfully recommend to your favorable consideration the application of the same system to this armory."

## AFTER THE WAR

Sept. 25, 1865, the following letter from Walter C. Hodgkins, formerly

Master Armorer of the C.S. Arsenal, Macon, Ga. written from New York City to Burton, late Colonel, C.S.A. in Macon, Ga., shows a friendliness and affection for his old comrade in arms:

"My dear sir:-

Yours of the 16th inst. came to hand this morning, and I was somewhat surprised to learn that you were still in Georgia.

My letters have been received very irregularly which I infer has been the cause.

I have been undecided (until recently) to write you as promised. I have now concluded to remain at least for the winter here, and have made business arrangements with Messrs. Cooper & Pond, 177 Broadway, where I will remain until something further develops itself.

I anticipate much pleasure in meeting you in person again. I have made enquiries concerning the Hotels in the neighborhood of the City Hall, and am compelled to admit that I do not think there is much choice. The "National" in Courtland St. a few doors from Broadway is kept on the European plan, the rooms are one dollar and one dollar and fifty cents each per day. The "Howard" on the corner of Maiden Lane and Broadway is three dollars and fifty cents per day and the "Powers" Hotel in Park Row at three dollars per day are both on the American plan and have been recommended, and if you could telegraph me from any point a day or so before your arrival, I think I could make you comfortable at either place. Before the war I had been stopping at the "Howard" for six or seven years, and always found the best of attention and the table and bed linen always clean.

I have several friends stopping at the "St. Dennis Hotel" corner of 11th St. and Broadway, and if it was not so far uptown I would say you would be better pleased there than at either of the other places, it is kept on the European plan and ranks with the best in the city. As your family is so large it will be better to endeavor to let me know the time of your arrival that apartments may be secured as most of the Houses keep pretty well full.

I have seen Col. Cuyler several times since I have been here. He was perfectly well, but undecided as to the future the last time I saw him (about two weeks ago). At one time he spoke of going to California and wanted me to accompany him but afterwards gave up the idea. I think he is stopping for the present in Amboy.

I have found many friends here, and almost every day meet some one of my Georgia friends together with my occupation make the time pass as pleasantly as I could expect under the circumstances.

Hoping you will make free to command me and that you will make your stay in New York as long as convenient, believe me

Yours sincerely

Walter C. Hodgkins"

The Cooper & Pond referred to in the above letter were firearms dealers at 177 Broadway, N.Y.C., from about 1862 to 1868. In 1868 the firm became Cooper, Harris & Hodgkins, and remained at the same address until 1876, when they removed to 298 Broadway. There they operated under the name of Hodgkins & Haigh, until they removed again in 1880, to 300 Broadway. This latter address had been occupied by W. J. Syms & Bro. during the war. Beginning with the city directory of 1887, Walter C. Hodgkins alone occupied 300 Broadway, and is still listed in 1888, but not later. Hodgkins's partner Haigh was a member of the firm of Onion Haigh & Cornwall in 1871, and this firm before the war operated as Onion & Wheelock. In 1871 they were at 18 Warren St., and later become Cornwall & Smoch, then Cornwall & Jesperson, then finally William Cornwall. When Hodgkins and Haigh moved to 298 Broadway in 1876, Edwin S. Harris stayed at 177 and was agent for Sharps. He sold out at auction Oct. 27, 1884, and never started up again—& neither did Spiller & Burr!

FINIS

**For Reference**

Not to be taken from this room

NORTH CAROLINA ROOM
NEW HANOVER COUNTY PUBLIC LIBRARY